# SIMPLE DIELECTRIC LIQUIDS

## MOBILITY, CONDUCTION, AND BREAKDOWN

BY

T. J. GALLAGHER

CLARENDON PRESS · OXFORD

1975

*Oxford University Press, Ely House, London W. I*

GLASGOW NEW YORK TORONTO MELBOURNE WELLINGTON
CAPE TOWN IBADAN NAIROBI DAR ES SALAAM LUSAKA ADDIS ABABA
DELHI BOMBAY CALCUTTA MADRAS KARACHI LAHORE DACCA
KUALA LUMPUR SINGAPORE HONG KONG TOKYO

ISBN 0 19 851933 8



Printed in Great Britain
by J. W. Arrowsmith Ltd., Bristol

# Preface

This book is an attempt to assess recent work on some of the important electrical properties of dielectric liquids. Another purpose is to try to reconcile many of the earlier conflicting results, especially those of breakdown measurements. This subject is now in a period of rapid growth, having been largely neglected in former years in favour of the more fashionable, and tractable, study of materials in the gaseous and solid states. Despite this neglect, there have been four international conferences on this topic in the past fifteen years, which have resulted in a vast amount of published work. There is now a need for a connected and critical account of the progress that has been made.

It is intended to cover only the research specifically concerned with simple dielectric liquids. An extensive account of work on complex liquids, such as mineral oils, by Zaky and Hawley was published recently. The emphasis throughout is on experimental results. Theoretical analyses are presented briefly but, where necessary, application of the various theories to results is discussed critically. This book contains a substantial amount of material which has not been collated previously. In particular, much of Chapter 1 is devoted to the mobility of fast charge carriers and to electrohydrodynamic (EHD) phenomena in liquids; recent models of conduction and breakdown can be found in Chapters 2 and 3, respectively; Chapter 3 also contains a discussion of the statistical interpretation of breakdown measurements with pulse voltages.

My thanks must go first to Professor J.J. Morrissey, who has encouraged me to continue this line of research, and provided the facilities. It is a pleasure to record my gratitude to Professor B.K.P. Scaife of Trinity College, Dublin, who read the typescript and made many helpful suggestions. My grateful appreciation is due also to my friends in this field who kindly furnished me with copies of their articles. I wish to acknowledge the help of my colleagues Professor E.A. McGennis, who took over part of my laboratory teaching, and Dr. A.J. Pearmain. I am especially indebted to Professor T.J. Lewis of University College, Bangor, who introduced me to this subject, and to Professor H. Fröhlich, F.R.S., University of Salford, who suggested the writing of this book. I also thank Mrs. R. Garvey who diligently typed the manuscript. Finally, a special word of thanks is due to my wife and children for their patience and endurance over the past fifteen months.

Acknowledgement is made to the Editors of the Annual Report of the Conference on Electrical Insulation and Dielectric Phenomena, British Journal of Applied Physics, Canadian Journal of Physics, Chemical Physics Letters, Direct Current, Institute of Electrical and Electronics Engineers Transactions on

Electrical Insulation, Journal of Applied Physics, Journal of Chemical Physics, Journal of the Electrochemical Society, Journal of Physics D: Applied Physics, Journal of Physical Chemistry, Nature, Physics of Fluids, Physics Letters, Physical Review, Physical Review Letters, Proceedings of the Institute of Electrical Engineers, Proceedings of the Physical Society, Proceedings of the Royal Society, Progress in Dielectrics; to the Librarian of the University of London; to the Directors of the Centre National de la Recherche Scientifique in Paris, the General Electric Research Laboratories in Schenectady and Taylor and Francis, London, for permission to use data from tables and to redraw figures.

*University College, Dublin*
*August 1974*

T.J. GALLAGHER

# Contents

# 1

# Mobility

## 1.1. Introduction

During the period of research which mainly concerns us in this book, namely the past 25 years, much of the early work emphasized the insulating, rather than the conducting, properties of dielectric liquids. Extensive measurements of the electric strengths of liquids were carried out but the results reported were often of a conflicting nature and a clear picture of the mechanisms of breakdown did not emerge from this work (Chapter 3). A vital clue, which could help to solve the secrets of breakdown, is hidden by the difficulty of identifying the main charge carriers involved in pre-breakdown conduction processes. Information about these carriers can be acquired through a measure of their mobilities. Consequently, in the last decade, considerable effort has been devoted to this type of measurement.

Interest in ion or electron charge carrier mobilities is not the sole preserve of those who have specialized in conduction and breakdown experiments in dielectric liquids. A knowledge of mobility is important for radiation chemists who wish to predict the kinetics of ion recombination and diffusion in liquids subjected to high-energy radiation. Nuclear physicists use dielectric liquids as the working media in particle counters and spark-chambers, and the operating speed of these devices is determined by the drift velocities of charges produced in the liquid. Furthermore, a proper interpretation of charge motion in fluids in terms of fundamental electron-atom, or ion-atom interactions can provide valuable information on the nature of the liquid state. Although our understanding of the mechanisms of conduction in insulating liquids still lags behind our comprehensive knowledge of the corresponding processes in the gaseous or solid phases nevertheless, in recent years, there has been substantial progress in the development of realistic theories to describe the charge transport properties of simple dielectric liquids. Mobility measurements have contributed significantly to this progress.

## 1.2. Methods of excitation and measurement

The 'natural' conductivity of dielectric liquids is generally very small and irregular. Therefore, to make the measurement of charge mobility easier it is necessary to enhance the normal charge density in a controlled manner, usually by some form of transient external excitation. The most common methods used are indicated in Fig. 1.1. In method (a) an $\alpha$-particle source, chemically deposited on one electrode, generates dense columns of positive ions and electrons along each $\alpha$-track. The range of the particle in liquids is $\sim 50\ \mu m$ (Aniansson 1955, 1961) so that the excess charges are created very close to the emitting electrode. With particle energies in the region of 5 MeV ($^{210}Po$) about $10^5$ ion pairs are generated per emission. However, because of recombination between the ion pairs, the number of electrons or positive ions escaping from each column is dependent on the applied electric field, a factor which can restrict mobility measurements to fields above $10^6\ V\ m^{-1}$, unless special techniques are used. An X-ray beam can be used to create positive and negative charges. This method has the advantage that the beam may be collimated and its energy varied to control the excess carrier densities and their place of generation. The bulk of the liquid may be irradiated or a very narrow section of it adjacent to one electrode, as shown in Fig. 1.1 (b). Electrons only can be injected into the liquid by ultraviolet (u.v.) illumination of a suitably chosen electrode, as in Fig. 1.1 (c), but care is needed to avoid charge generation in the body of the liquid itself. This problem is overcome by back-irradiation through a thin metal film evaporated onto a quartz window. However, in the latter case it is more advantageous to use electrons instead of u.v., as shown in Fig. 1.1 (d), since a thin metal layer may absorb almost all the light yet be practically transparent to an electron beam of several keV energy (Spear 1969). Furthermore, the density of electrons and their implantation depth in the liquid is easily controlled. Another method is to use a tunnel emitter of the metal–oxide–metal configuration, as in Fig. 1.1 (e). Under the bias conditions shown, 'hot' electrons, with temperatures equivalent to 15000K, can penetrate the thin film of $Al_2O_3$ into the Au electrode with extra kinetic energy, from which some of them are ejected into the liquid. Currents of, at least, 100pA can be emitted in a simple and controlled fashion and this technique promises to be very useful in examining the transport properties of charge carriers in liquids. Detailed instructions for the fabrication of these devices are available in the articles by Mentalecheta, Delacote, and Schott (1966), Savoye and Anderson (1967), and Onn, Smetjek, and Silver (1974). Finally, when a fine metal-point or edge electrode is raised to a sufficiently high negative or positive potential to produce a field near its tip of about $10^9\ V\ m^{-1}$, electrons or positive ions, respectively, can be created in the vicinity of the point (Fig. 1.1.f). Cold field extraction of electrons from metals in liquids has been studied in some detail in recent years, and indeed field emission from the cathode has long been considered as one of the primary causes of breakdown in liquids (Chapter 3). On the other hand, field ionization of liquid atoms or molecules to create positive ions has received little attention. However, its occurrence, as shown by the experiments of Halpern and Gomer (1965; 1969 *a, b*) and Schnabel and Schmidt (1973), does indicate that if the

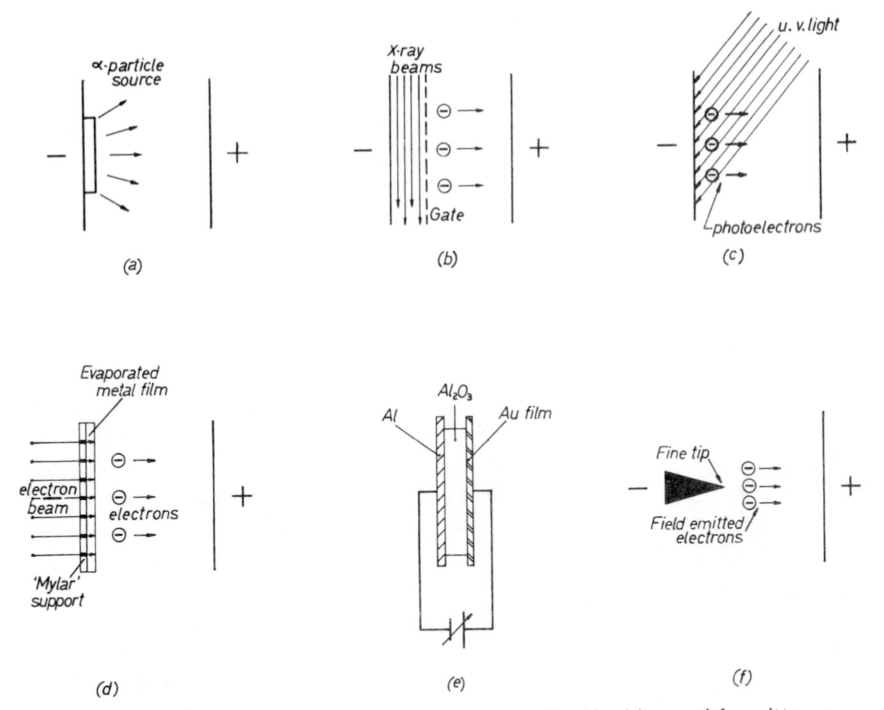


FIG. 1.1. Methods to enhance normal charge density in liquids: (a) α-particle emitter; (b) X-ray irradiation of a thin layer; (c) u.v. illumination of a photo-cathode, (d) electron beam; (e) tunnel emitter; and (f) field emission, or field ionization, from a fine metal point.

field at the positive electrode is large enough positive ions can be generated there. Field ionization, therefore, may also be an important factor in the mechanism of breakdown (section 3.5).

The mobility of a charge carrier is defined as its drift velocity per unit of electric stress. A direct estimate of mobility is determined by a time-of-flight method, which requires a measure of the time necessary for the charge to traverse a known distance in the liquid under the influence of a uniform electric field. Mobility values may also be inferred from a knowledge of the space-charge-limited currents for a given test-cell geometry. For detailed descriptions of the various experimental techniques of measurement the reader is referred to the extensive coverage given by Adamczewski (1969), and by Hummel and Schmidt (1971). A brief discussion of the basic principles is included here.

In its most simple form, the general arrangement for mobility measurements is illustrated by Fig. 1.2. Excess charge is created at the emitter electrode $E$ by any of the methods outlined in Fig. 1.1. By applying the appropriate polarity of voltage $V$ ions of one sign are swept to the collector electrode $C$. The motion of this ionic charge is manifest as a transient current in the external

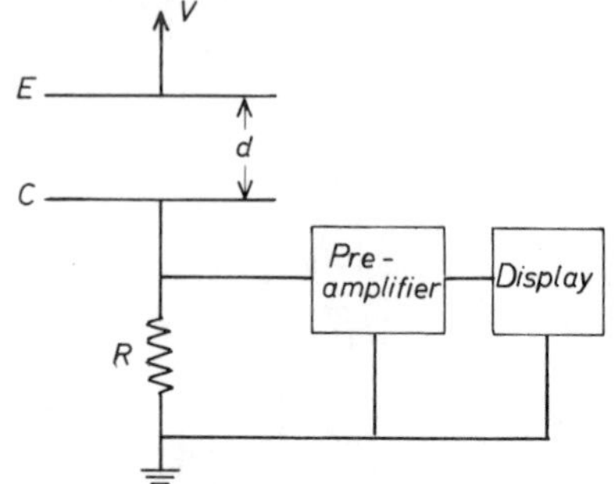


FIG. 1.2. Simple diode arrangement for mobility measurement.

detection circuit, and its arrival at $C$ is observed as a sudden change in the slope of the signal across $R$. The transit time $t$ for the carriers to drift a distance $d$ between $E$ and $C$ is measured from the oscillographic record of the signal across $R$ versus time, as shown in Fig. 1.3. The charge mobility $\mu$ is determined from the relationship

$$\mu = \frac{d^2}{tV}, \tag{1.1}$$

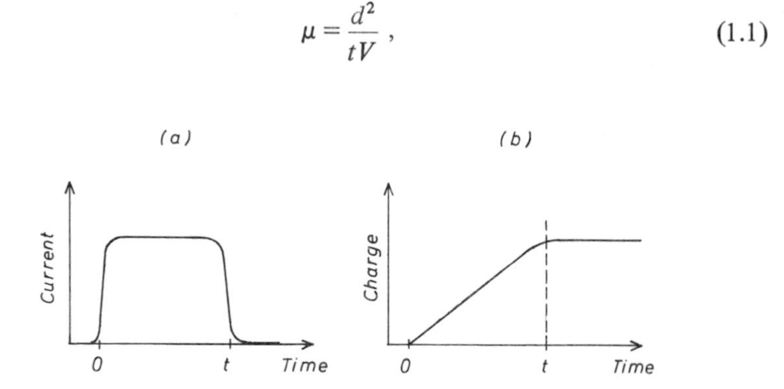


FIG. 1.3. Ideal signal shapes when carriers are created close to the emitter electrode: (a) current signal; (b) charge signal obtained from integration of the current signal.

The use of eqn (1.1) implies that the internal field in the liquid is not significantly perturbed by the drifting charge, and that it can be taken as the applied field. Accordingly, to avoid space charge distortion of the field the number density of the excess charge is limited to a small value ($10^{11}$ to $10^{13}$ $m^{-3}$), consistent with a measureable signal across $R$. In recent years the simple diode arrangement of Fig. 1.2 has usually been converted to triode or tetrode-type cells by the insertion of grid electrodes, suitably spaced between emitter and collector. Meyer and Reif (1958) were the first investigators to use grids for mobility experiments in insulating liquids. Two pairs were used as shown in Fig. 1.4 (a), and the technique of measurement was similar to that developed by Tyndall and Powell (1930) for ion studies in gases. Charge is generated by $^{210}$Po α-particles at *E*, which is maintained at a high potential relative to earth. The grid pairs *AB* and *DF* act as electrical shutters, or gates, to allow the passage of carriers across the drift space *BD*. The gates are arranged to open or close with the frequency of an a.c. voltage applied to them. If this frequency is changed continuously the number of ions reaching *C* is a maximum when their transit time between the gates is equal to, or an integral multiple of, the period of the pulses. The variation in the collector current is shown in Fig. 1.4 (b), where the

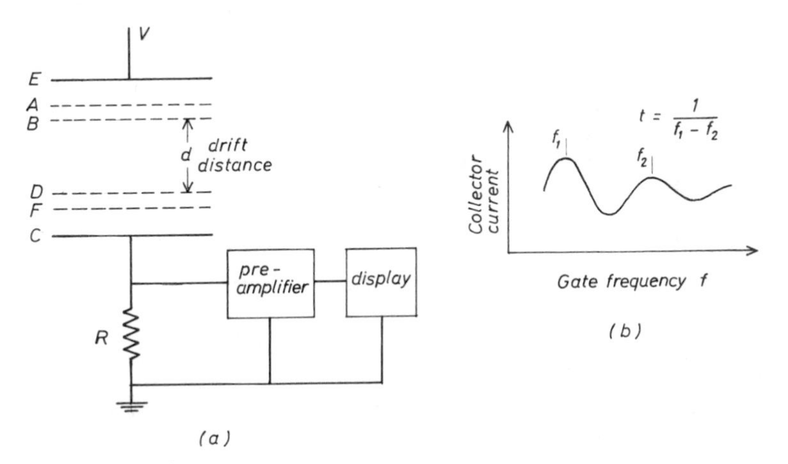


FIG. 1.4. (a) Diagram of double gate arrangement for mobility measurement, and (b) collector current as a function of the frequency of the voltage on the gates (after Meyer and Reif 1958).

transit time *t* is given by the reciprocal of the difference in frequency corresponding to adjacent current maxima. The amplitude of the oscillations tends to fall as the frequency of the gate voltage is raised with the result that the sensitivity of the method is decreased as the drift field is increased and the transit time shortened. The method has only been used with fields up to 25 $kVm^{-1}$ by Meyer and Reif (1958) working with liquid helium and by Schynders, Meyer, and Rice (1965, 1966) in liquid argon. Several variations on the circuit of Fig. 1.4 (a) have been developed (Cunsolo 1961, Bruschi, and Santini 1970) and cells containing one or two grids have gained widespread use in mobility studies. An upper limit to the applied field is now set by the mechanical stability of the finely-meshed grid or the risk of a discharge between the grid and the emitter. The use of gates permits extra control of the number of carriers extracted from the emitter region, provides better definition of the initial position of the injected charge, and improves the signal-to-noise ratio of the transient waveform.

## 1.3. Transport of slow charge carriers

The first extensive investigations of charge transport in liquids was made on organic fluids by Adamczewski (1937 *a, b*). In the interim, many papers have been published on this topic. Whilst there was a large disparity between the magnitudes obtained for mobility of either the positive or the negative carriers all investigators seemed to agree on one point—both species of ion behaved as slow-moving, singly-charged, carriers with molecular dimensions. Quite recently, there was a dramatic change in this situation. In 1969, both Minday, Schmidt, and Davis, and Schmidt and Allen, independently reported the detection, in n-hexane and tetramethylsilane, respectively, of quasifree electrons with mobilities four to five orders of magnitude greater than for negative ions. This discovery chronologically divides mobility measurements on organic liquids into 'slow' and 'fast' eras, and therefore, it is proposed to treat the motion of charge carriers under these separate headings. Measurements on cryogenic liquids will be treated in a similar fashion, although it is a long-established fact that the negative ion in liquid argon is a quasifree electron (Davidson and Larsh 1948).

### *1.3.1. Non-polar organic liquids*

For a number of non-polar hydrocarbon liquids, Table 1.3.1 gives a summary of the negative ($\mu_-$) and positive ($\mu_+$) ion mobilities which were obtained using time-of-flight techniques and three different methods of excitation. The other methods of carrier generation (Fig. 1.1) have been confined mostly to measurements on cryogenic liquids. There is a significant dispersion in the values in Table 1.3.1, expecially for $\mu_-$ in n-hexane, which is the most widely investigated liquid. There are several reasons for the popularity of n-hexane as a medium for investigation. It is liquid at room temperature, has a low viscosity which allows easy filtration, and has a convenient boiling point (341 K) for distillation and purification purposes. In addition, n-hexane, $C_6H_{14}$, is part of the homologous series of straight chain hydrocarbons, symbolized by the chemical formula $C_nH_{2n+2}$, so that any dependence of $\mu$ on molecular structure is readily examined by a simple alteration of chain length. For the same reasons hexane is widely used for conduction and breakdown studies.

A first glance at Table 1.3.1 would suggest that the spread in values for $\mu_-$ and $\mu_+$ is caused by the method of charge injection. This may partly be true since it has never been established that each form of excitation will produce an identical species of charge carrier. However, other factors such as sample preparation and induced motion in the liquid (sub-section 1.3.2) are mostly responsible for the scatter in results. We shall consider the results for $\mu_-$ first since, from a breakdown point-of-view, the behaviour of the negative carrier is of greater interest. Now, Minday *et al.* (1969) could only observe fast carriers in n-hexane after removing oxygen and other electron-scavenging impurities by extremely rigorous purification of the liquid. Consequently, the persistent

Table 1.3.1.

*Mobilities of slow charge carriers in hydrocarbon liquids*

| | Methods of excitation | | | | | | |
|---|---|---|---|---|---|---|---|
| | X-rays | | | Photoemission | | Field emission | |
| Liquid | $\mu_-$ | $\mu_+$ | ref. | $\mu_-$ | ref. | $\mu_-$ | ref. |
| n-Hexane | 4.4 | 8.27 | (1) | 11 | (10) | 4 → 14 | (15) |
| | 13 | 4 | (2) | 10 | (11) | 1.5 | (16) |
| | 9.2 | 5.8 | (3) | 10 | (12) | 1.5 | (17) |
| | | | | | | 6 → 8 | (17) |
| | 13 | 6.8 | (4) | 10 | (13) | | |
| | 9.1 | 8.5 | (5) | 7.5 | (14) | | |
| | | | | 1.5 | (14) | | |
| | 20 | 10 | (6) | | | | |
| | 2 → 2.5 | 2 → 2.5 | (7) | | | | |
| | 10 | 5 | (8) | | | | |
| | 6 | | (8) | | | | |
| | 2 | 2 | (8) | | | | |
| | 10 | - | (9) | | | | |
| n-Heptane | 6.6 | 4.2 | (3) | 11 | (10) | 2.4 | (16) |
| n-Octane | 5.2 | 2.9 | (3) | 7 | (12) | 3 | (16) |
| n-Nonane | 3.8 | 2 | (3) | - | | - | |
| n-Decane | 2.7 | 1.5 | (3) | 3 | (12) | 2.1 | (16) |
| Benzene | | | | 4.5 | (11) | | |

*Note*: In Table 1.3.1 mobility values are given in $(m^2 V^{-1} s^{-1}) \times 10^8$. References are identified by: (1) Adamczewski (1937); (2) Gzowski and Terlecki (1959); (3) Gzowski (1962*b*); (4) Hummel, Allen, and Watson (1966); (5) Schmidt (1968); (6) Secker and Lewis (1965); (7) Gray and Lewis (1969); (8) Belmont and Secker (1972); (9) Schmidt and Allen (1970*a*); (10) LeBlanc (1959); (11) Chong and Inuishi (1960); (12) Terlecki (1962); (13) Minday *et al.* (1971); (14) Brignell and Buttle (1971); (15) Essex and Secker (1968); (16) Essex and Secker (1969); (17) Taylor and House (1972*a*).

appearance of slow negative carriers in pre-1969 studies must mean that excess electrons, injected into 'impure' organic liquids, are rapidly converted into negative ions. The constitution of these ions is still a matter of speculation. Four distinct transformations have been suggested, namely that the injected electron is (*i*) attached to a neutral molecule as a negative ion, (*ii*) held in a partly trapped state from which it occasionally escapes, (*iii*) self-trapped in a cage of polarized liquid molecules as a type of polaron (von Hippel 1946) or (*iv*) scavenged by an impurity.

Because fast carriers are observed (*iii*) can be eliminated immediately. Model (*i*) is determined by the electron affinity of the neutral molecule and to the author's knowledge electron attachment, to form stable negative ions, has never been observed in non-polar hydrocarbons in the gas phase. It is most improbable, therefore, that electron attachment will occur to the molecules of any liquid listed in Table 1.3.1. LeBlanc (1959) has postulated model (*ii*) to explain the temperature dependence of mobility which he found for photoinjected electrons in n-hexane. As shown in Fig. 1.5, $\mu_-$ can be expressed by an Arrhenius-type relationship

$$\mu_- = \mu_0 \exp(-W/\mathbf{k}T) \tag{1.2}$$

where $\mu_0 = 3 \times 10^{-5}\ \mathrm{m^2 V^{-1} s^{-1}}$, and the activation energy $W = 0.14 \pm 0.02$ eV. LeBlanc also noted that the product of mobility and liquid viscosity $\eta$ changed with temperature (Fig. 1.5), indicating a violation of Walden's (1906) rule, which stated that the product of $\mu\eta$ should be a constant, independent of the nature of the solvent, and its temperature (Walden and Ulich 1923). The anomalous behaviour with respect to Walden's rule and the low mobility were taken as proof that the carrier was neither a free electron nor a negative ion, whereas the small value for $W$ was interpreted as the average activation energy of an electron in a shallow trap, having dimensions on a molecular scale.

Chong and Inuishi (1960), also using a photoemission technique, confirmed that $\mu_-\eta$ changed with temperature, and they obtained a value of 0.16 eV for $W$, in close agreement with LeBlanc. On the other hand, Gzowski (1962 *a, b*), using X-rays, verified Walden's rule for the negative ion in the series of n-paraffins hexane to decane. The positive species obeyed the relationship $\mu_+\eta^{\frac{3}{2}}$ = constant, in agreement with the earlier work of Adamczewski (1937 *a, b*). Similar findings were obtained for cyclohexane by Jachym (1963). The fact that results agree, or disagree, with Walden's rule appears to be over-emphasized in the literature. The rule cannot be considered quantitatively reliable for liquids with low relative permittivities (Robinson and Stokes 1965). Moreover, the rule is derivable from a form of Stokes's (1845) law:

$$v = \frac{F}{6\pi\eta R} \tag{1.3}$$

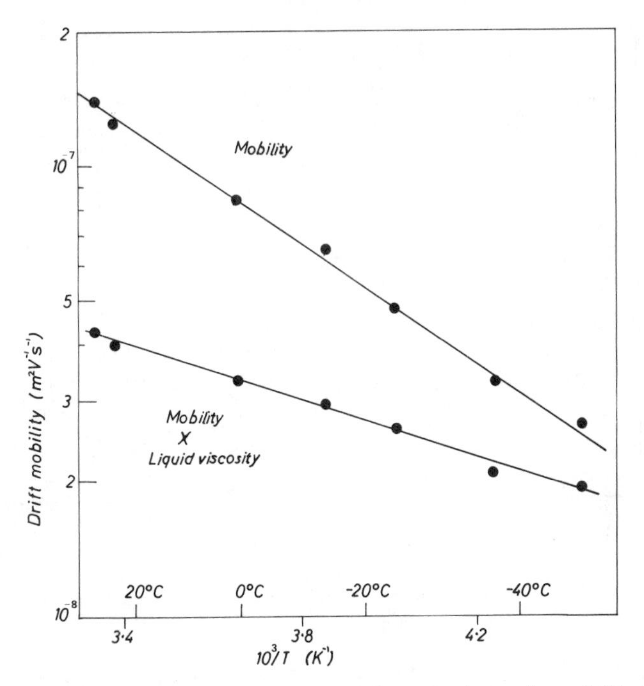


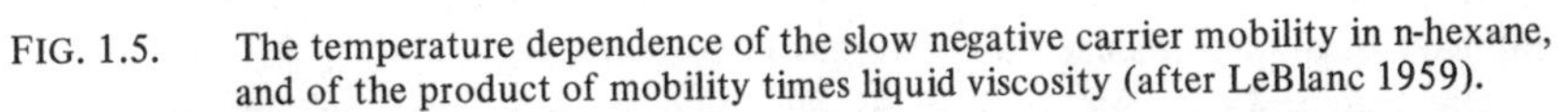
FIG. 1.5. The temperature dependence of the slow negative carrier mobility in n-hexane, and of the product of mobility times liquid viscosity (after LeBlanc 1959).

which relates the velocity of an uncharged spherical particle of radius $R$ to the driving force $F$ and the viscosity of the fluid through which it moves. Because of polarization forces the viscous drag on an ion may be very different from the frictional resistance experienced by an uncharged particle and whilst a Stokes–Walden relationship may give some idea of the size of ions it yields little information about their structures or their interaction with the host liquid.

LeBlanc's (1959) description (structure (*ii*) above) of the conduction state of a slow negative carrier in n-hexane is not supported by measurements of $\mu_-$ at high electric fields. Chong and Inuishi (1960) found that $\mu_-$ was independent of stress up to $5 \times 10^7$ V m$^{-1}$, the highest field at which direct mobilities have been recorded. Terlecki (1962) confirmed these findings in n-hexane, n-octane, and n-decane, as shown in Fig. 1.6. If the trapping model were valid then $\mu_-$ should increase rapidly at high fields because of a field-assisted reduction in the activation energy. This type of Poole–Frenkel effect is not observed (cf. Simmons 1967 for a treatment of this effect in solids). Consequently, model (*ii*) is rejected in favour of the remaining structure (*iv*), which involves the capture of electrons by impurity molecules. Further support for (*iv*) is provided by the observations of Schmidt and Allen (1970 *a*) who have shown that quasifree electrons are trapped to form conventional slow-moving carriers after the addition of impurities to ultra-pure liquids. Thus, some of the scatter of values in

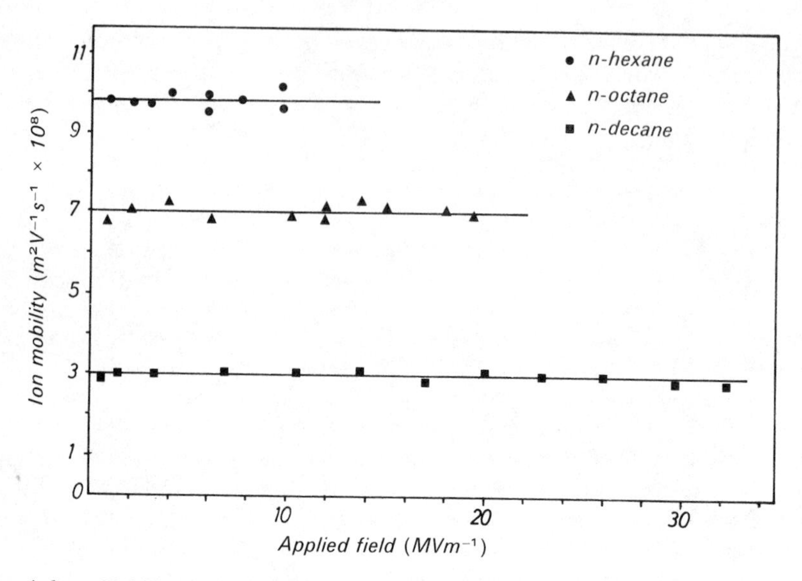


FIG. 1.6. Mobility of slow negative carriers at high applied fields in three n-alkanes (after Terlecki 1962).

Table 1.3.1 is easily reconciled with a variation in the amount and character of impurity in the liquids used by different investigators.

The principal impurity is probably dissolved oxygen. There is abundant evidence in the literature to show that in the gaseous phase oxygen can capture electrons to form negative ions of low mobility. Electron capture may occur by either of two mechanisms. Firstly, dissociative attachment to oxygen molecules is important at electron energies of several eV (Craggs, Thorburn, and Tozer 1957; Schulz 1962). This process can be expressed in reaction form by

$$O_2 + e^- \rightarrow (O_2^-)_{\text{unstable}} \rightarrow O + O^- + \text{energy} \tag{1.4}$$

Secondly, the molecular ion may lose its excess energy in a three-body collision to form a stable negative ion (Hurst and Bortner 1959). This is represented by

$$O_2 + e^- + X \rightleftarrows O_2^- + X + \text{energy} \tag{1.5}$$

and occurs at electron energies $< 1$eV. Foreign atoms and molecules may act as the third body X in stabilizing the attachment reaction in eqn (1.5), and because of the close packing of atoms or molecules in a liquid, three-body collisions should be very efficient. Swan (1963) has shown that the cross-section for electron attachment to oxygen in solution in liquid argon is appreciably greater than in pure oxygen in the gaseous state. Therefore, from the discussion in this sub-section, it would appear that the dominant negative carrier in 'impure' non-polar liquids is most likely an $O_2^-$ ion. The Oxygen probably attaches to itself by polarization forces a cluster of molecules to form a composite carrier during its migration through the liquids (De Groot, Gary, and Jarnagin 1967). For several n-alkanes Secker (1970) estimated that each cluster contained about nine molecules of the host liquid.

This model, which is essentially a picture of a solvated impurity ion, would explain the change in mobility shown in Table 1.3.1 for the series of 'impure' normal paraffins; since the extent of solvation, as well as the size of the solvating molecules, varies with the nature of the solvent, the effective ionic radius and mobility will not be constant. The simultaneous existence of two slow carriers in some experiments on n-hexane (see Table 1.3.1) probably arises through electron-scavenging compounds, apart from oxygen, which remain after liquid purification. It has not been possible to identify these impurities explicitly (Belmont and Secker 1971, 1972). Dissolved oxygen, as well as affecting the results of mobility measurements, can play an important role in determining the electric strength of liquids, as we shall see in section 3.7.

The transport of positive ions has not been examined to any large extent, probably because most investigators have visualized negative ions as the main charge carriers before, and during, breakdown. For all of the liquids in Table 1.3.1, and for many other organic liquids $\mu_+ < \mu_-$. As already noted, Gzowski (1962 *a*) found the relationship $\mu_+ \eta^{\frac{3}{2}}$ = constant yields a good fit to the experimental data, but, more recently, Gray and Lewis (1969) have stated that Walden's rule is obeyed for positive ions, provided corrections are made for the effects of liquid motion (sub-section 1.3.2). In view of our picture of the negative ion we can expect that the positive carrier also exists in a polarized atmosphere of neighbour molecules and it is this large entity that moves through the liquid.

### *1.3.2. Effects of induced liquid motion*

The mobility, or the drift velocity, of a charge carrier is an indication of the resistance to motion arising from collisions with the atoms or molecules of a liquid. Because of the many collisions suffered by each ion during its transit between emitter and collector sufficient momentum and energy may be exchanged to induce bulk motion of the liquid. The directed movement of ions will encourage liquid movement in the same direction, which will increase the ion velocity and produce an anomalously high mobility.

There is direct and indirect evidence for liquid motion. Ionic conduction can induce liquid flow sufficient to create hydrostatic pressures in the region of one tenth of an atmosphere. The phenomenon is known as ion drag pumping and has been comprehensively reviewed by Pickard (1965). It has been exploited in the electrical pumping of dielectric liquids by Stuetzer (1959, 1960), Kopylov

(1964), and Timko, Penney, and Osterle (1965). Stuetzer (1963) has also used ion drag pumping to analyse the electric field distribution in a liquid, whilst Blaisse, Goldschvartz, and Slagter (1970) have used it to develop a method of measuring ionic mobilities in liquid helium. The complementary effect, in which a moving liquid couples to charge carriers and gives them a directed motion, has been used by Lewa (1968), and by Kleinheins (1969 *a, b*) to study ionic mobilities in benzene and by Secker and Hughes (1969) in the construction of a Van de Graaff generator, using liquid n-hexane as the moving 'belt'. Direct evidence for liquid motion was obtained by Gray and Lewis (1965) who photographed the displacement of an uncharged droplet of dye introduced into a triode-type test cell containing hexane. With injection currents of $10^{-11}$ A from a $\beta$-particle source, and power inputs as low as $10^{-7}$ W, the liquid could attain a velocity in the region of $10^{-2}$ ms$^{-1}$, which was comparable with the estimated value for the drift velocity of the ions. In later experiments by Gray and Lewis (1969) in a tetrode cell, time-of-flight results were compared with displacement studies of the dye by recording the movement of the dye *after* switching-on the injection of carriers at the source gate. From a knowledge of the liquid motion the true carrier mobility with respect to the liquid was found. Fig. 1.7 shows the overall liquid and ion displacement–time curves. With a drift field of $10^5$ Vm$^{-1}$ the charge traversed the 15 mm gap in 2.4 s but, in this time, the liquid had also moved a distance of about $10.7 \times 10^{-3}$m. The ion displacement relative to the liquid

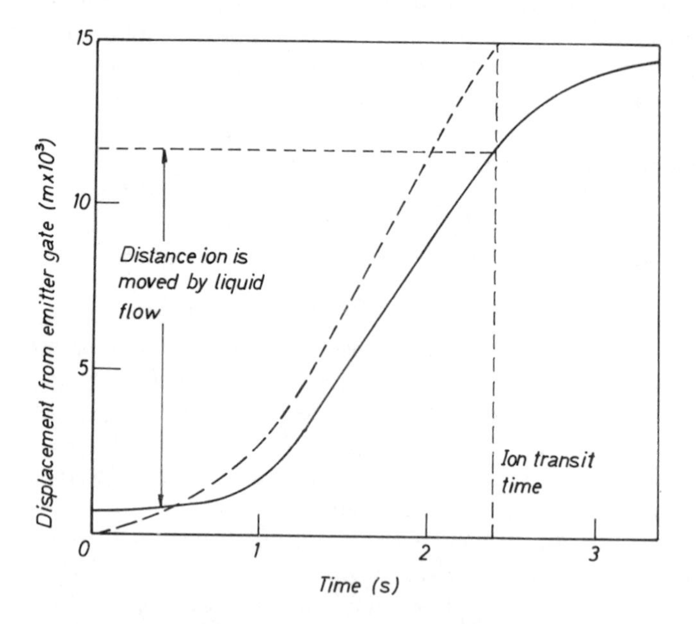


FIG. 1.7. Liquid and ion displacement–time curves. Full curve, liquid displacement; broken curve, ion displacement (after Gray and Lewis 1969).

was, therefore, only $4.3 \times 10^{-3}$ m. The true negative ion mobility is then $1.8 \times 10^{-8}$ m$^2$ V$^{-1}$ s$^{-1}$, whereas the apparent mobility is $6.25 \times 10^{-8}$ m$^2$ V$^{-1}$ s$^{-1}$. Important implications for mobility studies follow from these results. When $\mu$ is determined from a 'switch-on' transient, as in Fig. 1.7, by opening a gate to allow ions to enter the drift region, a certain time must elapse, because of inertia effects, before the liquid will begin to move. A corollary is that the apparent $\mu$ will be less than that calculated from a 'switch-off' transient, since liquid flow will already be established in the latter case. In fact Henson (1964), Secker and Lewis (1965), and Dey and Lewis (1968), amongst others, have observed that the 'switch-off' transient is shorter and more consistent than the 'switch-on' transient. Gray and Lewis (1969) have concluded that discrepancies between results of various investigators (Table 1.3.1) are explained by differences in the amount of liquid movement, brought about by the different injection currents and the heterogeneous design of the test cells used. By correcting for liquid motion Essex and Secker (1968) estimate the true $\mu_-$ to be $2.7 \times 10^{-8}$ m$^2$ V$^{-1}$ s$^{-1}$, in good agreement with Gray and Lewis, whilst Taylor and House (1972 *a*), using a laser doppler optical probe to eliminate errors due to liquid motion, quote a value of $1.5 \times 10^{-8}$ m$^2$ V$^{-1}$ s$^{-1}$ for $\mu_-$, which may still be larger than the true mobility because of inaccuracies in determining the liquid velocity.

Ostroumov (1954, 1956, 1962) appears to be the first person to have realized the significance of liquid motion in modifying the electrical properties of dielectric liquids. Ostroumov considered the electrohydrodynamic (EHD) stability of a liquid acted upon by an ion drag pressure gradient $qE$, arising from the presence of a charge density $q$ and applied stress $E$. From his analysis of several modes of flow, Ostroumov concluded that, in general, a liquid remains stationary at low stresses, but changes to a state of laminar motion at intermediate stresses, before turbulence sets in at higher stresses. The flow pattern for laminar EHD is illustrated by the simple vortex motion shown in Fig. 1.8, where the diameter of the vortices is comparable with the inter-electrode distance (Felici 1971 *a*). Turbulent EHD flow requires the creation of much smaller vortices. An instability in the system is said to occur when the bulk of the liquid is set in motion. Under experimental conditions there is no clear-cut transition from a stationary state to laminar and turbulent modes of flow. Ostroumov (1962) has shown that, in some cases, changes to the cell geometry can apparently induce a direct transition from a stationary to a turbulent state. The theoretical problems related to EHD instability have been re-examined in detail recently (see Note 1, Appendix), but a full discussion of the theories is beyond the scope of this book. Only the criteria for stability will be outlined as they give an indication of the conditions under which liquid motion can develop. Atten and Moreau (1969 *a*, *b*, 1970) have derived conditions for the onset of cellular motion in a parallel-plane electrode geometry, with injection of unipolar charge at one electrode, as in Fig. 1.8 above. The criteria are given in terms of non-dimensional numbers

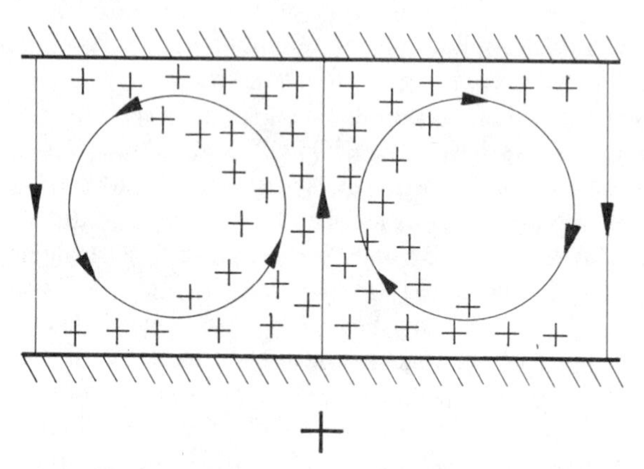

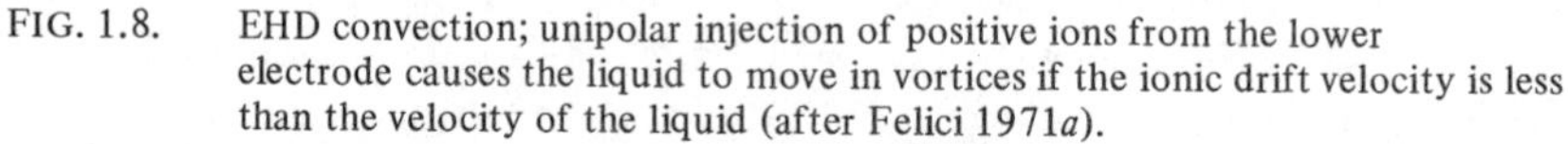

FIG. 1.8. EHD convection; unipolar injection of positive ions from the lower electrode causes the liquid to move in vortices if the ionic drift velocity is less than the velocity of the liquid (after Felici 1971*a*).

which are

$$M^2 C^2 R = 220.77 \quad \text{(weak injection)}$$

$$M^2 R \quad = 161 \qquad \text{(strong injection)} \tag{1.6}$$

where $C = \dfrac{q_0 d^2}{\epsilon V}$, $\mathrm{R} = \dfrac{\mu V \rho}{\eta}$, and $M = \dfrac{1}{\mu}\left(\dfrac{\epsilon}{\rho}\right)^{\frac{1}{2}}$,

$q_0$ represents the density of charge, with mobility $\mu$, at the injecting electrode, $d$ and $V$ the separation and potential between the electrodes, and $\epsilon$, $\rho$, $\eta$ are the permittivity, density and viscosity of the fluid respectively. For the case of strong injection, eqn (1.6) can be re-arranged to yield a critical voltage, $V_c = 161\mu\eta/\epsilon$, defining the stability criterion. $V_c$ is of the order of 300V for non-polar liquids and is in the region of 20V for polar liquids. Charge and stress are intrinsic to any mobility study so that convective motion of the liquid is likely to occur in most experiments. Schneider and Watson (1970) have studied convective flow arising from the injection of electrons at the free surface of a liquid and the limitation of current there by space charge. In this situation the criterion for stability is $M^2 R = 99$. This condition was also derived by Atten and Moreau (1972), who have extended their mathematical analysis to a consideration of the general case of an arbitrary finite injection at an electrode.

Experimental verification of the stability criteria is difficult. Because a high degree of purity and accurate control of charge injection are essential for experimental studies on EHD, polar, rather than non-polar, liquids have mostly been used. Brière and Felici (1960, 1964) developed purification techniques for polar liquids, using ion-exchange membranes and electrodialysis, which have provided the means for high levels of stable charge injection and have permitted controlled experiments in EHD. Filippini, Gosse, Lacroix, and Tobazeon (1969, 1970) and Filippini *et al.* (1970) have used Kerr cells coupled with a Schlieren optical system to map the propagation of the injected charge cloud and the associated wave-front, and to investigate the onset of instability. In nitrobenzene, $C_6H_5NO_2$, the critical voltage for an instability is between 2 to 16V compared with the theoretical prediction of $\sim$ 20V. Better agreement with theory was found by Lacroix and Tobazeon (1972) using a chlorinated diphenyl mixture (trade name Aroclor 1500). Nevertheless, experimental values of $V_c$ in polar liquids are consistently lower than predicted values. The reasons may lie in the neglect of parameters such as diffusion, electrolytic impurities etc., and in the use of semi-permeable membranes which are not ideal injectors (Atten and Moreau 1972). Electron bombardment of a free surface should permit a more uniform injection of carriers than is possible with membranes. Watson, Schneider, and Till (1970) used this method on a non-polar silicone liquid and obtained excellent agreement with their stability criterion.

From the previous discussion it is obvious that an error in results may be caused by liquid motion, and this possibility should be carefully assessed in a measurement of mobility. Unfortunately, the stability criteria in eqn (1.6) include the injected charge density $q_0$ which cannot be measured directly by the experimenter to provide him with a theoretical indication of the hydrodynamic state of the liquid. Hewish and Brignell (1972) have attempted to overcome this problem by deriving an expression for $q_0$ in terms of the experimental variables of current density $J$ and applied voltage $V$. Electrons were photo-injected into n-hexane and from measurements of $J$ and $V$, with known values for $d$, $\mu$, $\epsilon$, and $\eta$, the dimensionless numbers in eqn (1.6) were immediately evaluated with the aid of an on-line computer. Hewish and Brignell plotted their results in the form of a log $J$– log $V$ graph, as shown in Fig. 1.9, which also includes the space-charge-limited situation for a stationary liquid. From the graph it is apparent that the range of variables for which liquid movement is possible is very close to the space-charge-limited boundary. As this state of injection is avoided in time-of-flight measurements Hewish and Brignell have implied that previous determinations of mobility from transit times were not affected by liquid motion. They also point out that the results in Fig. 1.9 contradict earlier suggestions that either power or current density alone are adequate guides to the presence of an instability. It is difficult to reconcile these statements with the direct observations of fluid motion reported by Gray and Lewis (1969). However, we have already noted that the theoretical predictions of eqn (1.6) are always higher than experimental values, often by a significant amount. Consequently, to ensure a stationary liquid, it may be necessary to maintain test conditions far below the limits set by eqn (1.6). This

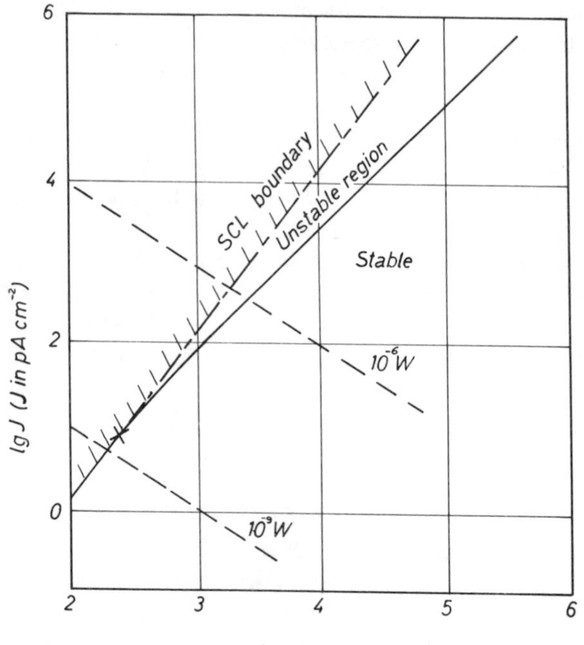


FIG. 1.9. The stability criterion for strong injection in eqn (1.6) plotted on a plane of lg $J$ against lg $V$. Also shown is the space-charge-limited boundary and lines of constant power input (after Hewish and Brignell 1972).

fact appears to have been overlooked by Hewish and Brignell (1972). Moreover, the density of bipolar ions resulting from X-ray or α-particle irradiation may produce a local environment in the liquid significantly different from that due to injected electrons. Thus the stability criteria in eqn (1.6) are not suited to experiments in which X-ray or α-particle methods of excitation are used.

*1.3.3. Polar liquids*

Polar materials such as water, acetone, ethanol, and to a lesser extent nitrobenzene, dissolve and dissociate impurities so efficiently that it is exceedingly difficult to purify them to a state where they exhibit high levels of resistivity. Traditionally, therefore, polar liquids have not been classed as insulators. Interest in their electrical properties was revived when methods of purification were developed which provided them with resistivities $\geqslant 10^{12}$ ohm-m (Brière and Rose 1967). This value is still small compared with $10^{16}$ ohm-m which is relatively easy to attain in many non-polar insulants. Nevertheless, if resistivities in the region of $10^{12}$ ohm-m can be maintained in them, highly-pure polar liquids are very attractive as dielectric media by virtue of their intrinsically large permittivities ($\epsilon_r > 30$) (see Note 2, Appendix). However, it is a most formidable task to maintain this ultra-pure state, since even glass can release enough impurity nearly to halve the resistivity. The mobility of charge carriers in polar fluids will also determine their usefulness as practical dielectrics. Although a few other liquids, e.g. acetone and sulpholane, can be purified to the same degree as nitrobenzene, almost all of the recent mobility studies in polar media have been confined to this liquid.

Brière and Gosse (1968) have measured $\mu_+$ in nitrobenzene by the simultaneous observation of the transient variation of the current and the field distribution in the liquid after a step voltage was applied. By means of a Kerr cell (see Note 3, Appendix), variations in the electric field were detected with relative ease in nitrobenzene, which has a Kerr constant that is much higher than for non-polar liquids. During the first 70 ms, Brière and Gosse (1968) found that $\mu_+$ depended on the purity of the sample and ranged from $2.5 \times 10^{-8}$ to $1.6 \times 10^{-8}$ m$^2$ V$^{-1}$ s$^{-1}$. When the sample purity was improved, $\mu_+$ increased to $2.3 \times 10^{-7}$ m$^2$ V$^{-1}$ s$^{-1}$, a value which was significantly higher than other previous estimates. In the period from 200 ms to 1 min, $\mu_+$ varied between $10^{-8}$ and $10^{-9}$ m$^2$ V$^{-1}$ s$^{-1}$ as determined from the field distribution near the electrodes. It was assumed that this field was due to a layer of space charge, and Silver (1965) has shown that $\mu$ can be estimated from the thickness of this layer and the voltage drop across it. The experiments of Brière and Gosse (1968) illustrate well a major problem encountered with polar liquids; results are influenced by the amount of impurities, particularly of an ionic nature, remaining in the liquid. Atten and Gosse (1968) achieved unipolar injection into very pure nitrobenzene with ion-exchange membranes in contact with the electrodes. By covering the cathode with an anodic membrane protons or potassium ions were released into the liquid. The technique was reversed when chlorine or picrate negative ions were studied. An identical mobility of $2 \times 10^{-7}$ m$^2$ V$^{-1}$ s$^{-1}$ was measured for all of these ionic carriers, which is an order of magnitude greater than their true mobilities as determined from conductimetry experiments. Atten and Gosse (1968) attributed these high mobilities to EHD effects, and this was confirmed by Filippini *et al.* (1969), and Hopfinger and Gosse (1971), who monitored the induced motion in nitrobenzene using Schlieren (see Note 4, Appendix) techniques (see Plate 1). By relating charge mobility to the time for filaments of liquid to cross the inter-electrode spacing an 'EHD mobility' was obtained, in excellent agreement with the high value of Atten and Gosse (1968).

Little is known about the nature of the negative ion in impure polar liquids. The major charge carrier will depend on the residual ionic impurity level but 'free' electron conduction is precluded because of the low values of mobility. For pure nitrobenzene Bright, Makin, and Pearmain (1969) have suggested that the negative carrier is probably the radical anion $C_6H_5NO_2^-$, formed by electron attachment to the molecule. Weiss (1960) postulated a similar mechanism in water. Felici (1967) has proposed that polaron-like structures are created by clusters of molecules surrounding the anion radical, and that charge transport occurs via the transfer of electrons between the clusters. Even less is known about the type of positive ion but Felici (1971 *a*) has surmised that protons, either from water or absorbed hydrogen, are the main transporters of positive charge in polar media.

### *1.3.4. Cryogenic liquids*

Literally, the term 'cryogenic' means icy cold, but it is now common practice to restrict this description to liquids which are produced at temperatures less than approximately 200 K. This classification includes the liquid forms of He, Ne, Ar, Kr, and Xe, and the diatomic gases $H_2$, $O_2$, and $N_2$. The cryogens are also aptly termed liquefied gases. These liquids exhibit relatively simple structures and a fair degree of local order (Henshaw, Hurst, and Pope 1953) because only the weak van der Waals bonds are responsible for the cohesive forces between their constituent atoms or molecules. Consequently, they are of great scientific interest as they should behave as prototype models as far as the structure of liquids and the conducting states of disordered systems are concerned. Interest in the insulating properties of these low-temperature liquids has also flourished in recent years. The absence of discharge products makes them eminently suited to a fundamental investigation of dielectric breakdown. Indeed, the results of breakdown studies has demonstrated the remarkable influence that the anode as well as the cathode can exert on the electric strength of liquids (section 3.5). Furthermore, cryogenic liquids are now being used in many engineering applications such as compact rocket fuels, as cryopumps for large vacuum systems, in superconducting magnets and machines, and, obviously, as refrigerants.

Before discussing the transport of slow charge carriers in cryogenic liquids it is necessary to become acquainted with the nomenclature of liquid helium. Atmospheric helium gas consists of a mixture of two isotopes $^4$He and $^3$He in relative concentrations of $10^6 : 1$. We shall confine our attention to liquid $^4$He since liquid $^3$He is difficult to produce and there are very few investigations of its electrical properties (cf. McClintock 1973). Unlike other substances $^4$He does not have a triple point, but, on either side of a transition temperature, the liquid exists in two modifications. The transition occurs at the lambda-point temperature ($T_\lambda$) of 2.176 K. The 'high' temperature form of $^4$He is called HeI and, in general, it may be considered as a classical liquid. The low temperature form HeII has many strange properties, being a mixture of two fluids, the 'normal' component which has a definite viscosity, and the 'superfluid' component which has zero viscosity and zero entropy. HeII is designated a quantum liquid. For a comprehensive account of the physical properties of liquid helium reference should be made to the book by Wilks (1967).

The natural conductivities of cryogenic liquids are negligibly small. Therefore, in order to study the mobility of ions and electrons in these liquids, it is necessary to generate carriers artificially; α-, or β-particle sources are the most usual methods of excitation but any of the techniques of Fig. 1.1 are suitable. For example, Halpern and Gomer (1965, 1969 *a*), Blaisse *et al.* (1970), and McClintock (1969, 1973) used field emission, Miller, Howe, and Spear (1968) used electron beams, and Smetjek, Silver, Dy, and Onn (1973) used tunnel diodes to inject excess carriers into liquefied gases.

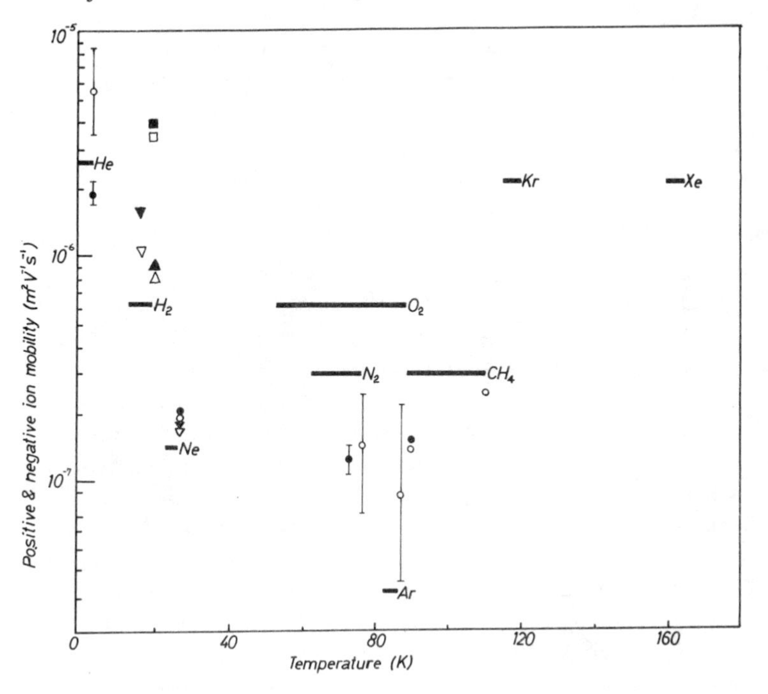


FIG. 1.10. ▬, the temperature range of some cryogenic liquids. Open and closed symbols refer to $\mu_+$ and $\mu_-$, respectively. $H_2$ : ▾▿ Gachechiladze *et al.* (1970); ▪▫ Halpern and Gomer (1969*a*,*b*); ▴▵ Zessoules *et al.* (1963). Ne: ○● Lebedenko and Rodionov (1972); ▿▾ Loveland *et al.* (1972*b*). $O_2$ : Loveland *et al.* (1972*a*). See text for references to work on other liquids.

Fig. 1.10 is an attempt to show the temperature range of cryogenic liquids at normal vapour pressures, as well as the range of measured mobilities for both species of ionic carrier, at temperatures near their boiling points. The wide range of mobilities, in some of the liquids, is probably caused by EHD effects (subsection 1.3.2). Several other features of Fig. 1.10, however, require special comment. Firstly, data are not shown for the parent negative ion in Ar, Kr, and Xe. Whilst slow negative ions have been detected in these liquids by numerous investigators, it is now accepted that they are probably in the form of $O_2^-$ impurity ions and that ions such as $Ar^-$ etc. do not exist. We shall see in section 1.4 that excess electrons can attain velocities in Ar, Kr, and Xe nearly $10^4$ times faster than these impurity ions. Secondly, in He at 4.2 K, $\mu_+$ is at least three times greater than $\mu_-$. This situation is rare in dielectric liquids and raises interesting questions regarding the structure of the ions in He. We shall consider possible structures later in this sub-section. Another fascinating aspect of charge motion in these liquids is the reports by some investigators of small, but definite, reductions in ionic mobilities, which appear at critical values of their drift velocities. Figs. 1.11 and 1.12 show this behaviour for $\mu_-$ in HeII and $\mu_+$ in Ar, respectively. Henson (1970) has observed seven distinct values for $\mu_+$ in HeI whilst Bruschi, Mazzi, and Santini (1970) found changes in $\mu_+$ and $\mu_-$ in HeI, $N_2$, and Ar, and in carbon tetrachloride at room temperature. Huang and Olinto (1965) have postulated that in HeII the ions gather excess energy from the electric field to create quantized vortex rings (Rayfield and Reif 1964) at certain critical velocities, whilst Henson (1964) has proposed that the step changes in Ar and $N_2$ are caused by changes in the effective hard-sphere cross-sections of ion clusters, which alter in size at critical electric fields. However, these theories are probably superfluous as far as discontinuities in $\mu$ are concerned. A back-to-back arrangement of two tests cells was used by Goodstien, Buontempo, and Cerdonio (1968) to show that the critical velocity is constant for vortex ring formation in HeII, and that the Huang–Olinto theory is not correct. Schwarz (1970, 1972) and Steingart and Glaberson (1970) have made deliberate, and very accurate but unsuccessful, searches for the discontinuities in HeII, as have Doake and Gribbon (1971) in $N_2$. In addition, many investigators of ionic mobilities have not detected any step-changes in their results.

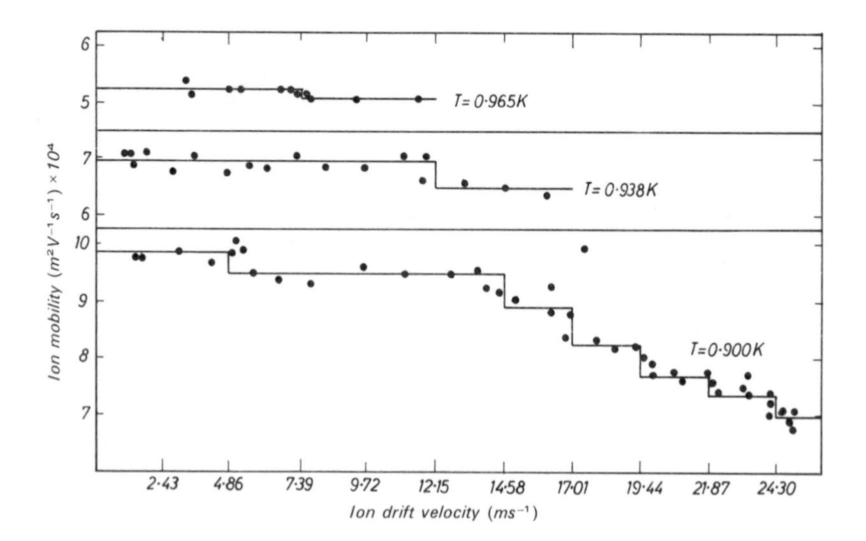


FIG. 1.11. Step changes in $\mu_-$ in liquid helium II (after Careri, Cunsolo and Mazzoldi 1964).

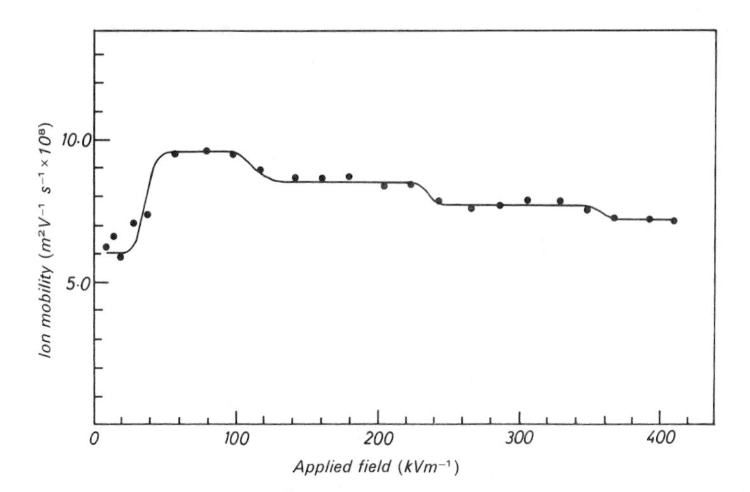


FIG. 1.12. Step changes in $\mu_+$ in liquid argon (after Henson 1964).

To explain the phenomena Doake and Gribbon (1971) adopted an earlier suggestion by Dey and Lewis (1968) that the appearance of discontinuities in $\mu$ is intimately connected with induced liquid motion. This would appear to be the most satisfactory explanation, especially in view of our remarks on EHD effects in sub-section 1.3.2. With the exception of He the changes in $\mu$ occur at critical drift velocities in the region of $10^{-2}$ m s$^{-1}$. This is comparable with the induced velocity of n-hexane measured by Gray and Lewis (1969), and strongly suggests that the discontinuities are of a hydrodynamic origin. In fact, Bruschi, Mazzoldi, and Santini (1966, 1967) have noted that the critical ionic velocity in HeII is dependent on the hydrodynamics of the normal and superfluid components of the liquid. Moreover, those investigators reporting the existence of jumps have measured $\mu$ from the 'switch-off' transient (sub-section 1.3.2) and, therefore, adequate time was probably available for liquid motion to be generated. There is one factor which appears to contradict these hydrodynamic theories; jumps in $\mu$ imply corresponding changes in the hydrodynamic impedance of the test cells used. Such abrupt changes will not happen. Nevertheless, Fisher and Taylor (1971) and Taylor and House (1972 *b*) have observed that the magnitude of the current and its pattern of flow were modified significantly by the interelectrode hydrodynamic conditions. It may be significant that Schwarz (1972), and Steingart and Glaberson (1970) used drift spaces of $27.13 \times 10^{-2}$ m and $8.7 \times 10^{-2}$ m respectively, which were much larger than in other experiments. Any change in the resistance to liquid motion in these larger test cells may not be reflected by strong perturbations in the pattern of flow and, consequently, jumps in $\mu$ would not be evident in their experiments. The controversy may be resolved by making mobility measurements in a cell where the drift space can be varied between short and long distances.

Several models of ionic structure have been mooted for the slow charge carriers in liquefied gases. Four possibilities exist for the negative ion, namely: (*i*) an ion of the form $A_n^-$ where A symbolizes the element in question, (*ii*) a charged impurity ion, (*iii*) an electron localized in a cavity, and self-trapped in a shell of polarized atoms, or (*iv*) a quasifree electron. For the rare-gas liquids, (*i*) can be discarded immediately, since ions of the type $He_n^-$ or $Ar_n^-$ are unstable, if they are formed at all. This statement also applies to $N_2^-$, and the slow negative ion detected by Bruschi *et al.* (1970), and by Loveland, Le Comber, and Spear (1972 *a*), was probably an $O_2^-$ impurity ion. The work of Swan (1963), and Dey and Lewis (1968), has shown that oxygen concentrations as low as 1 p.p.m. in Ar can trap many injected electrons to form $O_2^-$ ions with greatly reduced mobility, whilst Davis, Rice, and Meyer (1962 *a*) attributed their low carrier mobilities in Ar, Kr, and Xe to the transport of impurity oxygen ions. We shall see in section 1.4 that model (*iv*) describes the state of excess electrons in pure Ar, Kr, and Xe. The solubility of all substances in liquid helium is very small and, in any case, impurities are probably frozen out onto the walls of the test cell. Model (*iii*), therefore, is the only form which can account for the structure of the negative ion in He. A similar model for negative ions in Ne has been postulated by Miyakawa and Dexter (1969), Loveland *et al.* (1972 *b*), Bruschi *et al.* (1972), and Lebedenko and Rodionov (1972), and for negative ions in $H_2$ by Gachechiladze, Mezhov-Deglin, and Shal'nikov (1970).

At a first glance it is rather surprising that excess electrons do not remain 'free' in liquid helium, as they appear to do in other rare gas liquids such as argon, krypton, and xenon. The low value for $\mu_-$ in helium is taken as an indication that there is a strong interaction between electrons and helium atoms in the liquid phase. A theoretical treatment of the electron–He atom interaction by Jortner, Kestner, Rice, and Cohen (1965) has shown that it is of a short-range repulsive form, with a magnitude of the order of several eV. Evidently, then, the electron creates a cavity around itself by radial repulsion of neighbouring helium atoms. Thus, the electron exists in a bubble and, in fact, several theoretical treatments of this model have considered that the electron is simply confined inside a square potential well. The bubble model was first considered on a theoretical basis by Kuper (1959) who, however, attributed the original idea to R.P. Feynman. Kuper (1959) showed that the electron is localized within a sphere of radius 15 Å and has an effective mass equivalent to about $10^2$ helium atoms. Since then many extensive experimental and theoretical studies have been published about this model for the negative ion in liquid helium. The bubble dimensions estimated by Kuper are regarded as of the correct order.

Sommer (1964) obtained experimental evidence in favour of the repulsive nature of the electron–He atom interaction. The free surface of liquid $^4$He was bombarded with electrons from a source situated in the vapour phase, and the currents that flowed along the surface and through the bulk liquid were measured. From the results, Sommer (1964) estimated, with an accuracy of about ± 30 per cent, that electrons with energies less than about 1.3 eV could not penetrate the surface energy barrier. Recently, Schoepe and Rayfield (1973) examined the escape of electrons through the surface of liquid He. Electrons were believed to tunnel from their bubble states and the height of the surface potential barrier was calculated as 0.82 eV. The type of electron-atom interaction may also be determined by comparing the effective work function of a photocathode immersed in a liquid with its work function in a vacuum. The difference for any metal surface is considered to represent the energy of the ground state of an electron in the liquid. The difference is denoted by $V_0$, where

$$V_0 = \Delta\phi = \phi_{\text{liquid}} - \phi_{\text{vacuum}}. \tag{1.7}$$

As shown in Fig. 1.13 positive values of $V_0$ were obtained by Woolf and Rayfield (1965), who measured the spectral response of a cesium–antimony cathode in HeII and in vacuum at 1.1 K. Woolf and Rayfield (1965) interpreted the upward shift in work-function to mean that the electron favours a bound, localized, energy state, typical of a bubble structure. $e\Delta\phi$ from Fig. 1.13 was estimated to be + 1.02 eV, in good agreement with Sommer's (1964) experimental value. We should note, in passing, that Halpern, Lekner, Rice and Gomer (1967) measured a downward shift of 0.33 eV in liquid argon, as shown in Fig. 1.14. Halpern *et al.* (1967) interpreted this negative value for $e\Delta\phi$ as an indication that the energy state of the electron in Ar is delocalized and that it can exist as a quasifree particle. Mobility measurements in argon seem to confirm this picture of the electron (section 1.4). Levine and Sanders (1962, 1967) obtained the most direct experimental evidence for the validity of a bubble model of the negative ion in He. These authors measured the electron mobility as a function of the density of He gas and they observed that $\mu_-$ dropped by a factor of $10^3$ to $10^4$ when a critical density of the gas at 4.2 K was reached. The decrease in $\mu_-$ was explained by the electron undergoing a transition from a delocalized, energy state, typical of a bubble structure. $e\Delta\phi$ from Fig. 1.13 was state as the helium was compressed from gaseous to liquid densities. Northby and Sanders (1967) have apparently succeeded in ejecting electrons from their bubble states in HeII by irradiating the negative ions with infrared light. The results imply a bubble radius of 21 Å and a 'square-well' depth of 1 eV. Meyer and Reif (1961) and Springett (1967), have examined the pressure dependence of $\mu_-$ in HeII, which has provided another test of the bubble model. Rather surprisingly, it was observed that $\mu_-$ at first increased to a maximum for pressures up to about 7 bar, but, thereafter, $\mu_-$ decreased in a linear fashion for pressures up to about 20 bar. Nevertheless, by treating the negative ion as an electron trapped in a potential well with 'soft' sides, as in a bubble, Springett

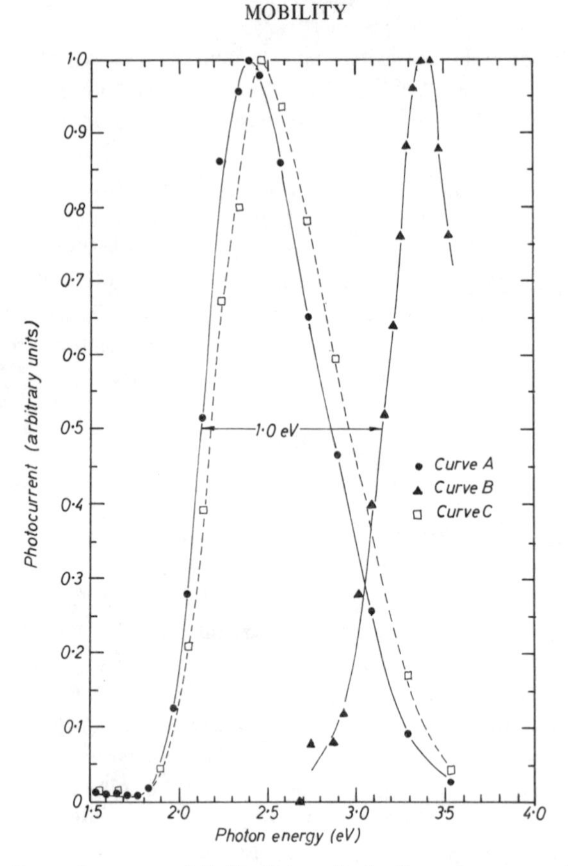


FIG. 1.13. Spectral response of Cs-Sb photo cathode. Curve A, vacuum at 1.1K; curve B, liquid helium II at 1.1K; curve C, helium gas at a pressure of $10^{-5}$ torr (after Woolf and Rayfield 1965).

(1967) has accounted for the pressure dependence of $\mu_-$, and has estimated that the bubble radius decreased from 16 Å to 10 Å as the pressure was increased from 0 to 20 bar. From all of the discussion in this sub-section we must conclude that a bubble model is the correct description of negative ions in liquid helium. Also, as we have seen, this model is generally assumed to apply to negative ions in liquid hydrogen and liquid neon. However, this assumption need not be correct as will be shown by reference to some observations at the end of sub-section 1.4.1. It is not possible in a book of this size to delve more deeply into the prodigious volume of literature on excess electrons in He which has been published in the last two decades. For a more extensive treatment of this topic the reader is referred to the excellent article by Rice (1968). Additional references are cited in a literature survey by Gauster and Schwenterley (1973), and in a brief review of the electrical properties of low-temperature liquids by Gallagher (1973).

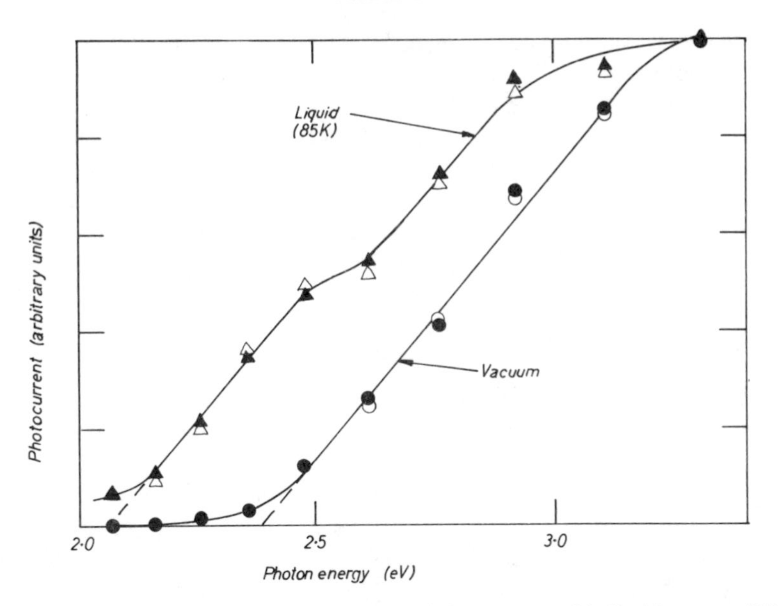


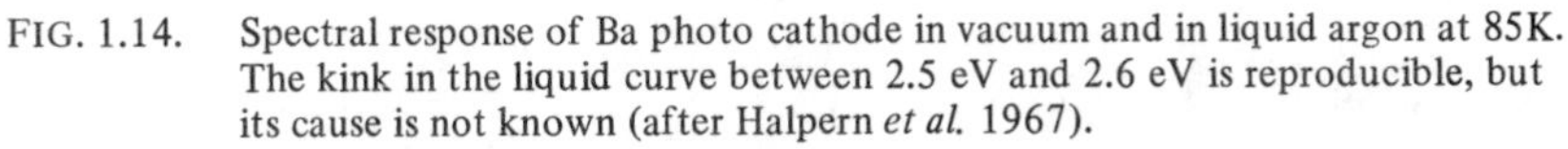
FIG. 1.14. Spectral response of Ba photo cathode in vacuum and in liquid argon at 85K. The kink in the liquid curve between 2.5 eV and 2.6 eV is reproducible, but its cause is not known (after Halpern *et al.* 1967).

Compared with the detailed investigations of the character of negative ions in cryogenic liquids there has been little interest in the structure of the positive charge carriers. Nevertheless, it is now generally accepted that the stable forms of the positive carriers are the diatomic ions $He_2^+$, $Ar_2^+$, etc. Bakale and Schmidt (1973 *a*) have suggested that the positive species in liquid methane is probably a $CH_4^+$ ion. Significant polarization around each ion is likely, as in the case of positive ions in organic liquids. Indeed, for liquid helium Atkins (1959) has calculated that the polarization forces may create electrostrictive pressures greater than 25 bar. These pressures are sufficient to solidify the helium atoms surrounding the ions and, thus, the ion is transported as a high-density particle. Atkins (1959) derived a radius of 3.5 Å for the ion, which contained about 50 He atoms. This radius is in good agreement with the 4.3 Å estimated by Meyer, Davis, Rice, and Donnelly (1962) using the Stokes Law eqn (1.3). However, the most accurate value of the positive ion radius is probably $5.0 \pm 0.1$ Å, obtained by Schwarz and Stark (1969), for HeII between 1 K and 0.426 K. Because of the superfluid properties of HeII in this range of temperature, $\mu_+$ varies by nearly four orders of magnitude. Nevertheless, Schwarz and Stark (1969) accounted for this variation simply by treating the mobility in terms of collisions between a 'hard-sphere' ion and the elementary roton excitations of the liquid. The Rice and Allnatt (1961) kinetic theory of transport phenomena in classical fluids has been applied by Davis *et al.* (1962 *b*) to their measurements of $\mu_+$ in liquid argon. The measured value of $6.1 \times 10^{-8}$ m$^2$ V$^{-1}$ s$^{-1}$

is in excellent agreement with the predicted value of $5.93 \times 10^{-8}$ m$^2$ V$^{-1}$ s$^{-1}$. However, Dey and Lewis (1968) have measured a lower value of $3.6 \times 10^{-8}$ m$^2$ V$^{-1}$ s$^{-1}$ for $\mu_+$ which, undoubtedly, was augmented by EHD effects. Consequently, Dey and Lewis (1968) have suggested that the true $\mu_+$ for a stationary liquid is probably about $2 \times 10^{-8}$ m$^2$ V$^{-1}$ s$^{-1}$, and that the ion-atom interaction potential employed by Davis *et al.* (1962 *b*) in the Rice–Allnatt theory needs to be modified. Rice and Jortner (1965) have discussed in great detail the kinetic theories of dense fluids and their application to ionic transport in liquids.

## 1.4. Transport of fast carriers

### *1.4.1. Fast carriers in cryogenic liquids*

Fast negative ions were observed first in liquid argon. Davidson and Larsh (1948, 1950), Hutchinson (1948), and Marshall (1954) used the experimental arrangement of Fig. 1.2 to investigate liquid argon as a possible dielectric medium for a high-energy particle detector. These authors concluded that Ar was a suitable liquid because the conductivity pulses produced by $\alpha$-particle irradiation showed that the excess negative carriers possessed drift velocities in the region of $10^3$ to $10^4$ m s$^{-1}$. These values were approximately a factor of $10^5$ greater than the known velocities of heavy negative ions in liquids. Consequently, the negative charge carrier in liquid argon was considered to be a quasifree electron.

Using the particle detector technique, with improvements to the electronic circuitry, Malkin and Schultz (1951), Williams (1957), Swan (1962 *a*, 1964), and Pruett and Broida (1967), attempted direct measurements of the electron drift velocities in liquid argon. The results of these investigators are summarized in Fig. 1.15, which shows the variation of drift velocity $v$ over a range of electric fields $E$, between 1 MV m$^{-1}$ to 10 MV m$^{-1}$. Results for fields $< 1$ MV m$^{-1}$ are not possible with the technique used, because of the limitations already mentioned in section 1.2. The saturation of $v$ at high $E$ in Fig. 1.14 is a common trend in all the measurements. However, the disparity in the absolute magnitude of $v$ is outside the limits of experimental error, especially as similar methods were used by each investigator. EHD effects will not significantly influence the transport of fast carriers, since their drift velocities are so much greater than induced liquid velocities. Dissolved impurities are responsible for the discrepancies in Fig. 1.15 according to Swan (1964), who has shown that the addition of about 0.2 to 0.4 per cent of gaseous $H_2$ or $N_2$ to liquid argon effectively doubled the drift velocity. Molecular impurities influence $v$ in an identical fashion in the gaseous state of the rare gases. The impurity effect in gaseous argon is quantitatively accounted for by the energy dependence of the elastic collision cross-section $Q$ of argon atoms to electrons. $Q$, which is known as the Ramsauer cross-section (see Note 5, Appendix), is found to decrease rapidly as the mean electron energy is reduced from about 10 eV to 1 eV. In effect, the

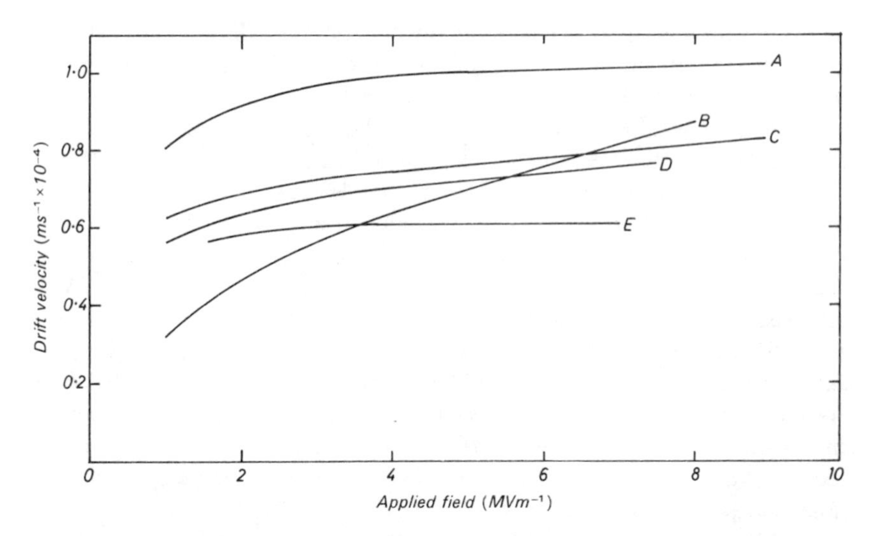


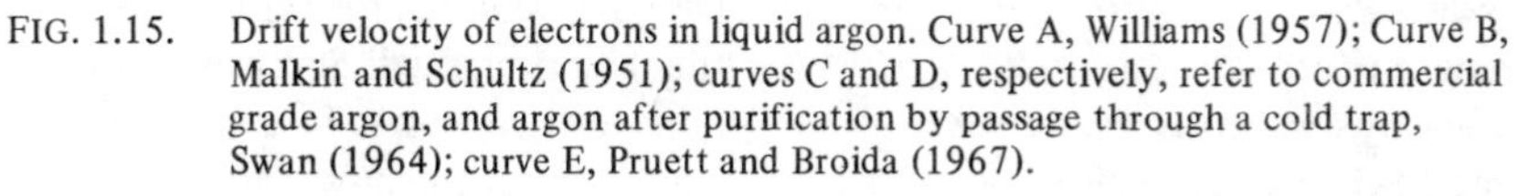
FIG. 1.15. Drift velocity of electrons in liquid argon. Curve A, Williams (1957); Curve B, Malkin and Schultz (1951); curves C and D, respectively, refer to commercial grade argon, and argon after purification by passage through a cold trap, Swan (1964); curve E, Pruett and Broida (1967).

argon atoms become more transparent to electrons as the energy of the electrons is reduced. Traces of molecular impurities such as $N_2$, which introduce inelastic rotational and vibrational collisions in the energy range from $10^{-3}$ eV to 1 eV, will effectively reduce both the cross-section and the mean electron energy. Hence, $\nu$ is increased, since it is approximately inversely proportional to the cross-section. Swan (1964) has interpreted impurity effects on $\nu$ in liquid argon using the arguments above, although he realized that the interpretation neglected any influence of the liquid structure on the motion of the electrons. Moreover, Pruett and Broida (1967) have suggested that inelastic scattering by impurities is responsible for the saturation in drift velocities in Fig. 1.15. However, calculations by Lekner (1967), which take account of effects due to the structure of the liquid, have indicated that a Ramsauer cross-section does not exist in liquid argon.

The reader will have noticed that so far in this sub-section we have omitted any reference to electron *mobilities*. The reasons for this omission are obvious from a glance at Fig. 1.15. Since the drift velocity is practically field-independent no ratio of $\nu$ to $E$ can yield a meaningful value of mobility. The ratio will yield an apparent mobility for which physical explanations can be deduced in a qualitative sense only.

In order to assess in a quantitative manner the available theories of transport phenomena in liquids, it is essential to obtain drift velocity data at low applied fields. At low values of stress, the electrons are in thermal equilibrium with the host liquid because the energy gained by the electrons from the field is small compared with their thermal energy. In this situation the distribution of electron energies is Maxwellian, and the Boltzmann (see Note 6, Appendix) transport equation, with suitable modifications to account for structural effects, can be used to derive the drift velocity of electrons in liquids. At high electric stresses, the field energy of the electrons is greater than their thermal energy, and they are no longer in equilibrium with the liquid. In this situation the energy distribution of the electrons is characterized by an effective temperature which is greater than the liquid temperature. In effect, the distribution is 'heated' and the electrons are classed as hot electrons. Cohen and Lekner (1967) have solved the transport equation for hot electrons in gases, liquids, and solids. Schynders *et al.* (1965, 1966) were the first investigators to measure the velocities of quasifree electrons in liquid argon and liquid krypton at low applied fields. From the point of view of liquid purity this seemed to be a most formidable experiment. Earlier measurements by Swan (1963) of the electron attachment coefficient of impurities in liquid argon had indicated that an $O_2$ impurity concentration as low as 1 part in $10^9$ was necessary before electrons could be detected at low fields. Schynders *et al.* (1965, 1966) removed oxygen by passing the argon gas through activated charcoal. After liquefaction, the liquid was further purified by electrolysis at a stress of 10 kV $m^{-1}$ for two hours. This procedure enabled Schynders *et al.* to obtain a $v$–$E$ graph in argon which was linear up to 20 kV $m^{-1}$, the highest stress used. The slope of this graph, at 85K, gave an electron mobility $\mu_e$ of $4.3 \times 10^{-2}$ $m^2$ $V^{-1}$ $s^{-1}$. A linear extrapolation of this measurement to 1 MV $m^{-1}$ yields a value of $v$ nearly five times greater than the highest value for this stress in Fig. 1.15. It would appear, therefore, that electrons migrating in liquid argon are scattered by different mechanisms at low and high fields. The electron beam technique of Fig. 1.1.d was developed by Miller *et al.* (1968) in order to extend measurements of $v$ from low to high fields in Ar, Kr, and Xe. As shown in Fig. 1.16, the findings of Miller *et al.* (1968) confirm the results of Halpern *et al.* (1967), and of Swan (1964). The data of Schynders *et al.* (1965, 1966) are in agreement with the results at low fields in Fig. 1.16, as are the recent findings of Robinson and Freeman (1973 *a*). The field dependence of $v$ obtained by Miller *et al.* (1968) in Kr and Xe followed the pattern of Fig. 1.16.

The $v$–$E$ curve in Fig. 1.16 can be divided into three regions. First, a low-field, ohmic, region where $v$ is linearly dependent on $E$ and the ratio $v$ to $E$ yields a constant value of electron mobility. Secondly, at about 20 kV $m^{-1}$ there is a transition to a region where $v$ exhibits an $E^{\frac{1}{2}}$ dependence. It is intriguing to note that Miller *et al.* (1968) observed this transition both in the liquid and solid states of Ar, Kr, and Xe when $v$ was near the velocity of sound in each medium. A similar shape in the $v$–$E$ curve has been obtained by Bakale and Schmidt (1973 *a, b*) for electrons in the liquids methane and neopentane. In each case the transition to region 2 occurred when $v$ was near five times the sonic velocity. Consequently, in this region it would appear that $v$ is determined

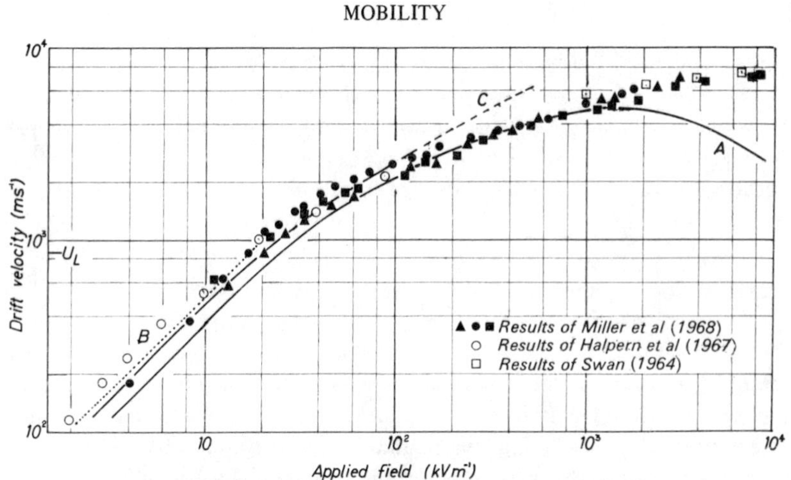


FIG. 1.16. Drift velocity of electrons in liquid argon at low and high applied fields (after Miller *et al.* 1968).
Curve A, calculated from Cohen–Lekner theory; Curve B is the Cohen–Lekner theory corrected for multiple scattering (Halpern *et al.* 1967). Curve C is calculated from the theory of hot electrons in solid germanium (Shockley 1951). $U_L$ is the velocity of sound in liquid argon.

by phonon scattering whereby electrons lose their energies to acoustical modes of vibration in liquids. Thirdly, in Fig. 1.16, there is a high field region where $v$ tends towards saturation, as in the measurements of Fig. 1.15. The non-linear field dependence of $v$ above a stress of 20 kV $m^{-1}$ is indicative of a heating up of the electron distribution, and Miller *et al.* (1968) interpreted their results on the basis of the hot electron theory of Cohen and Lekner (1967). By including the effects of liquid structure in their solution of the Boltzmann equation, the latter authors made a significant advance in the analysis of electron motion in liquids. The theory of Cohen and Lekner (1967) can be used to estimate drift velocities if the effective atomic potential for electron scattering in the liquid is known. This potential is intimately related to the structure of the liquid and Lekner (1967) made detailed calculations, specifically for liquid argon. His results for the scattering potential show that a Ramsauer cross-section does not exist in liquid argon at temperatures in the region of 85 K. Curve A in Fig. 1.16 represents the $v$–$E$ curve calculated by Lekner (1967). The agreement with experiment is excellent for fields up to approximately 2 MV $m^{-1}$. At higher fields the theory predicts a maximum in the $v$–$E$ curve rather than the saturation which is typical of all the experimental data. Lekner (1967) has suggested two reasons for this discrepancy; either the theoretical scattering cross-sections are inaccurate at high fields or the experimental results are influenced by impurities. Miller *et al.* (1968) have discounted impurity effects and have concluded that the former reasons, or possible approximations in the theory, are responsible. It is of interest to note that Lekner (1967) estimated a mean energy of 2.4 eV for electrons in liquid argon at a stress of 10 MV $m^{-1}$. On this basis it is possible that some hot electrons in the high-energy tail of the distribution could reach ionizing energies. Ionizing collisions, at a stress of 100 MV $m^{-1}$, have been postulated as a possible cause of dielectric breakdown in liquid argon (section 3.5).

For low applied fields, Lekner (1967) gave the electron drift velocity as:

$$v = \left(\frac{2}{3}\right)\left(\frac{2}{\pi m \mathbf{k} T}\right)^{\frac{1}{2}} \frac{eE}{n\,4\pi a^2\,S}\quad, \tag{1.8}$$

where $e$ and $m$ are the electron charge and mass, respectively, $n$ is the number density of the liquid, **k** is Boltzmann's constant, $T$ is the absolute temperature, $a$ is the electron–atom scattering length, and $S$ is the liquid structure factor. The latter is related to the isothermal compressibility $\chi_t$ by $S = n\mathbf{k}T\chi_t$. According to eqn (1.8), the low field mobility should show a $T^{-\frac{3}{2}}$ variation with temperature. This behaviour was found for temperatures up to about 120 K by Jahnke, Meyer, and Rice (1971), who extended the measurements of Schynders *et al.* (1966) to temperatures and pressures from the triple point to the critical point of liquid argon. Above 120 K some extraordinary results were obtained. Fig. 1.17 shows a representative curve of the apparent electron mobility as a function of temperature, at a constant pressure of 50 bar. A similar pattern was observed on other isobars, with the maxima occurring at a common number density in the liquid of approximately $1.2 \times 10^{28}$ $m^{-3}$. Also shown in Fig. 1.17 is the remarkable correlation between the electron mobility and the bulk viscosity of liquid argon, which has been noted by Freeman (1973). The latter has also suggested that mobility and viscosity measurements might be used as complementary methods to probe the structure of liquids. Lekner (1968 *a, b*) has tried to account for the anomalous high temperature, low density behaviour of $\mu_e$ in Fig. 1.17 but, at present, there are no satisfactory explanations. Several ideas for a fresh theoretical analysis of the problem have been suggested by Jahnke, Holzwarth, and Rice (1972).

Before concluding our discussion of fast carriers in cryogenic liquids, it is necessary to mention two results which appear to have escaped attention in the literature: both Malkin and Schultz (1951) and Pruett and Broida (1967), have hinted at the presence of quasifree electrons in their measurements on liquid neon. These observations contradict the picture of a bubble form for the negative ion in neon (sub-section 1.3.4). Furthermore, the experiments by Pruett and Broida (1967) in some of the solidified rare gases showed that the quasifree electron velocities systematically decreased in the order Ne, Ar, Kr, and Xe. Velocities in the solid state were only about twice as large as in the liquid state of Ar and Kr, the only substances which Pruett and Broida examined, in detail, in both states. Similar results were obtained by Miller *et al.* (1968), who studied

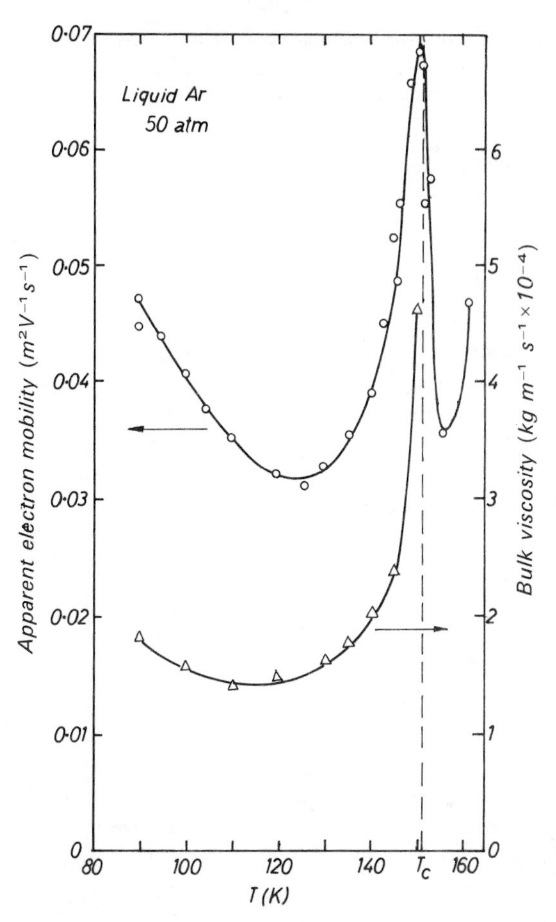


FIG. 1.17. Correlation between the apparent electron mobility in argon and the bulk viscosity of the liquid (after Freeman 1973).

the two phases of Ar, Kr, and Xe. On the other hand, Keshishev, Mezhov-Deglin, and Shal'nikov (1970), and Dionne, Young, and Tomizuka (1972), have shown that excess electrons introduced into solid He migrate as negative ions with velocities some $10^3$ times slower than in liquid He. Electrons would be expected to have a large mobility in the more highly ordered solid, whereas ions will be almost immobile in the solid. As electron–atom interactions and quantum effects in neon are somewhat similar to those in helium, the latter result would suggest that a corresponding trend in drift velocities should occur for the two phases of neon. Experimentally, an opposite trend has been observed. Clearly then, if a bubble model is to represent the real nature of the negative ion in liquid neon at the change of phase from solid to liquid, the electron velocity would have to decrease by a factor of approximately $10^5$. It is difficult to imagine such a drastic reduction taking place. It may be that the slow negative ions, previously reported in liquid neon, were connected with the presence of impurities. On the other hand, quasifree electrons may migrate in solid helium. The measurement circuits used to examine solid helium had response times much too slow to detect fast carriers. Obviously, there is much scope for further investigation of the conducting states of electrons in liquid neon and in the solid phases of neon and helium.

### *1.4.2. Fast carriers in hydrocarbon liquids*

As mentioned in section 1.3.1, it was only in 1969 that the first brief reports were published of fast negative charge carriers in organic liquids. In n-hexane Minday *et al.* observed a mobility at least 50 times larger than previous values whilst Schmidt and Allen measured a mobility $10^5$ times greater in tetramethylsilane. Tewari and Freeman (1968), and Conrad and Silverman (1969) had inferred the presence of these fast carriers from the observations of high mobility transients during pulse radiolysis of hydrocarbons.

In order to detect these fast carriers extremely elaborate purification of the liquids was required. To remove impurities Schmidt and Allen (1970 *a*) employed a variety of chemical agents as well as sodium–potassium mirrors, molecular sieves, silica gel, and repeated vacuum distillation techniques. The major procedure used by Minday *et al.* (1971) involved prolonged vacuum pumping on the liquids, and repeated exposure of them to barium film getters. Barium films chemisorb $H_2$, $N_2$, $O_2$, $H_2O$, CO, and $CO_2$ but do not affect hydrocarbons. Some 30 to 100 cycles of barium gettering and pumping were needed before a species of fast carrier was detected. Minday *et al.* (1971) injected excess electrons into their liquids by photoemission from a barium-coated nickel cathode, whilst Schmidt and Allen (1970 *a*) used either thin layer, or bulk liquid, X-ray irradiation to create carriers of both signs.

Fig. 1.18 shows the results of Minday *et al.* (1971) for n-hexane, and some unsaturated and aromatic liquids. It is seen that the drift velocities of the negative carriers are much greater than for negative ions in n-hexane, yet are several orders of magnitude lower than for quasifree electrons in liquid argon. Minday *et al.* (1971) verified in several ways that the velocities were, in fact, related to the transport of electrons. By measuring the photocurrents transmitted through the liquid–vapour interface for a sample containing fast carriers and slow negative ions, it was observed that the fast carriers penetrated the surface barrier into the vapour. This would be expected of electrons only. Also, a reversal of the polarity of the field between emitter and collector decreased the collector current by at least two orders of magnitude, indicating that the fast carriers must have originated at the cathode. Furthermore, identical results were obtained using either a simple diode, or a double gate, test cell arrangement. If the carriers were large entities of an ionic nature, some difference in results would be expected, even from EHD effects alone (sub-section 1.3.2). Notwithstanding

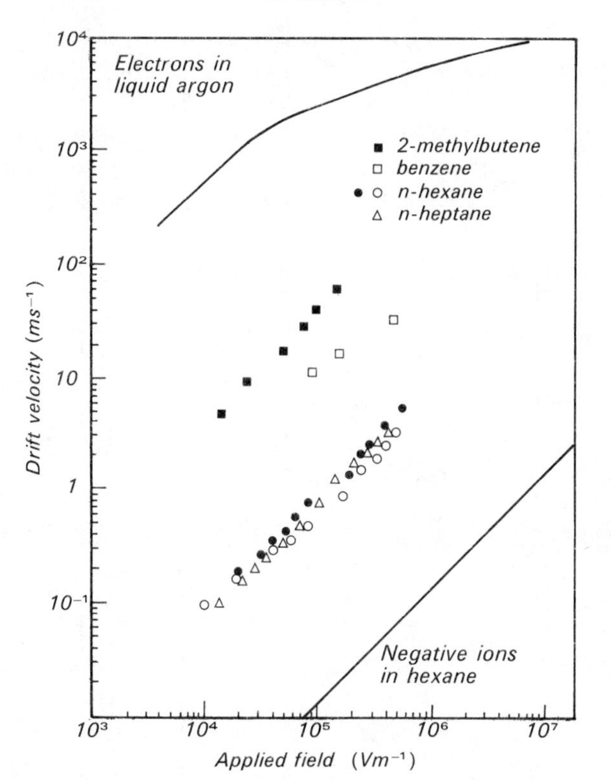


FIG. 1.18. Electron drift velocity versus applied field in several hydrocarbon liquids (after Minday *et al.* 1972).

that the fast carriers are electrons, the results of Fig. 1.18 indicate that hydrocarbons such as n-hexane offer much more effective resistance to electron transport than the heavy liquefied rare gases. Obviously, different mechanisms for electron scattering must operate in the two classes of fluid. We shall discuss the mechanisms in organic liquids later in this sub-section.

Electron mobilities in paraffin hydrocarbons and in a few other liquids are listed in Table 1.4.1. The list is restricted almost entirely to liquids which are discussed in Chapter 3 in relation to breakdown measurements. Freeman and his colleagues have inferred electron mobilities for a vast range of liquids (see Note 7, Appendix) from the electron conductance transient produced during pulse radiolysis. With reference to Table 1.4.1 there is substantial agreement between the values for $\mu_e$ in n-hexane obtained by Schmidt and Allen (1970 *a*) and by Minday *et al.* (1971). However, there is a factor of 2, at least, between their values for n-pentane. Residual impurities in the liquid would appear to be the only explanation for this difference. Rather than trap electrons the impurities may alter the scattering cross-section and, consequently, the mobilities, as happens with nitrogen in liquid argon (sub-section 1.4.1). Schmidt and Allen

Table 1.4.1.
*Electron mobilities in various dielectric liquids*

| | Liquid | Structural formula | Mobility $(m^2 V^{-1} s^{-1}) \times 10^4$ | *bd* $(kgm^{-2}) \times 10^7$ | Temperature (K) |
|---|---|---|---|---|---|
| (1) | n-Hexane | $CH_3(CH_2)_4CH_3$ | 0.09 | 44.2 | 296 |
| | n-Pentane | $CH_3(CH_2)_3CH_3$ | 0.16 | 44.6 | " |
| | n-Butane | $CH_3(CH_2)_2CH_3$ | 0.40 | 47.7 | " |
| | 2,2,4-Trimethylpentane (iso-octane) | $(CH_3)_2CHCH_2C(CH_3)_3$ | 7 | 65.4 | " |
| | 2,2-Dimethylbutane (neohexane) | $C(CH_3)_2{=}CH(CH_3)_2$ | 10 | 59.5 | " |
| | Neopentane | $C(CH_3)_4$ | 55 | 104.9 | " |
| | Tetramethylsilane | $Si(CH_3)_4$ | 90 | 103 | " |
| | Cyclopentane | $CH_2-CH_2 > CH_2$, $CH_2-CH_2$ | 1.1 | 51.1 | " |
| | Cyclohexane | $C_6H_{14}$ | 0.35 | 51.3 | " |
| (2) | n-Hexane | – | 0.07 | – | 300 |
| | n-Pentane | – | 0.07 | – | " |
| | Benzene | $C_6H_6$ | 0.60 | – | " |
| | 2-Methylbutene-2 | $(CH_3)_2C{=}CHCH_3$ | 3.60 | – | " |
| (3) | Methane | $CH_4$ | 400 | – | 111 |
| | Ethane | $(CH_3)_2$ | 0.80 | – | 200 |
| | Neopentane | – | 70 | – | 295 |
| | Neohexane | – | 12 | – | " |
| (4) | Methane | – | 530 | 224 | 111 |
| | Ethane | – | 0.97 | 57 | 200 |
| | Propane | $CH_3\ CH_2\ CH_3$ | 0.55 | 52 | 238 |
| | n-Butane | – | 0.27 | 54 | 293 |
| | Cyclopropane | $(CH_3)_3$ | 0.011 | 39 | 260 |

(1) Schmidt and Allen (1970 *a, b*); (2) Minday *et al.* (1971); (3) Bakale and Schmidt (1973 *a, b*);
(4) Robinson and Freeman (1974).

(1970 *a*) have reported that in all of their liquids the drift velocity exhibited an ohmic behaviour up to the highest fields used by them. This field was 10 MV $m^{-1}$ in n-hexane and 1.5 MV $m^{-1}$ in neopentane. However, according to Bakale and Schmidt (1973 *b*), the difference in the two values for $\mu_e$ in neopentane in Table 1.4.1 is because the lower result of Schmidt and Allen (1971 *a*) relates to an apparent mobility which was determined from the drift velocities of hot electrons. Nevertheless, we can conclude from the results of Schmidt and Allen (1970 *a*) that electrons in polyatomic liquids remain at thermal energy even at relatively high fields. This behaviour is in marked contrast to the non-linear field dependence above 20 kV $m^{-1}$ for Ar, Kr, and Xe in Fig. 1.16, and it reflects the greater efficiency of energy transfer between electrons and polyatomic molecules by inelastic collisions.

The most striking aspect of all of the results in Table 1.4.1 is that electrons have widely different mobilities in those hydrocarbons which have very similar chemical and physical properties. For example, $\mu_e$ in neopentane is greater than in n-pentane by a factor of almost 300 although the densities of the two liquids are comparable. More remarkable still is the tremendous increase in $\mu_e$ from ethane to methane. In methane an electron moves almost as fast as in liquid argon. In passing, it is apposite to remark that a methane molecule resembles an argon atom in that both are spherical and they offer almost identical Ramsauer cross-sections to electrons in the gaseous phase. It would be interesting to see if drift velocity measurements in methane, with trace amounts of gaseous $N_2$ added, would yield results similar to those found in an $Ar-N_2$ system (subsection 1.4.1). The results in Table 1.4.1 also show a marked correlation with $\mu_e$ and the spherical symmetry of the molecules. On the one hand, high mobilities are found in neopentane and tetramethylsilane (TMS), which possess nearly spherical molecules. On the other hand, the long chain asymmetrical molecules in n-hexane and in n-pentane give rise to low mobilities. Molecular sphericity is also important on another account. When molecules are ionized by irradiation the electrons which escape from their parent positive ions will dissipate their large excess energies over some distance before they reach thermal energy. This thermalization distance, denoted by $b$ in the literature, will determine the yield of free ions from irradiated liquids. Tewari and Freeman (1968, 1969), Fuochi and Freeman (1972), and Schmidt and Allen (1968, 1970 *b*), amongst others, have observed that $b$ increased with increasing molecular symmetry. As might be expected, there is a striking correlation between $\mu_e$ and $b$. Values of $b$ vary from 70 Å in hexane and pentane, through 214 Å in neopentane and TMS, to 592 Å in methane and 1330 Å in argon. The penetration depth of the electron should be inversely proportional to the density $d$ of any material. Consequently, the density-normalized product $bd$ should be more characteristic of the degree of interaction and energy loss of electrons in the liquid. Values for $bd$ have been included in Table 1.4.1. Again, there is a good correspondence between $\mu_e$, $bd$, and the sphericity of the molecules. Dodelet and Freeman (1972) have attributed the influence of molecular structure on *bd* to the anistropy of polarizability of the molecules. Molecular excitation processes are related to the anisotropy of polarizability. Because methane and ethane have widely different values of *bd* (see Table 1.4.1), Dodelet and Freeman (1972) concluded that Raman-type rotational excitation (see Note 8, Appendix), rather than intra-molecular vibrations, act as the principal mechanism for the dissipation of electron energies in hydrocarbons. Energies in the region of $10^{-2}$ eV are associated with these rotational modes, whereas at least $10^{-1}$ eV is required to excite the various modes of vibration. The influence of molecular structure on the electric strength of liquids has been attributed, by some investigators, to the presence of vibrational, rather than rotational, excitations between constituent groupings in each hydrocarbon molecule (sub-section 3.9.1).

It is still necessary to account for the enormous range of $\mu_e$ in liquid hydro-carbons. The range is bounded by the low mobility for a localized bubble state of the electron, as in helium, and by the high mobility for a delocalized quasifree electron, as in argon. Thus, to explain the continuous variation in $\mu_e$ in hydrocarbons,it is reasonable to picture the electron as occasionally being trapped during its migration through a liquid. The extent of trapping will be governed by the scattering potentials of the different molecules. However, the nature of these traps is not well understood. Schmidt and Allen (1970 *b*) have pictured them as a 'fortuitous coincidence of rotational phases of neighbouring molecules'. As such, they may be regarded as shallow traps arising from collective molecular arrangements in the liquids. These traps must not be con-fused with the process of electron trapping by chemical impurities. The traps are intrinsic to the liquid itself. Essentially, they arise from voids in the fluid where the electrons may reside for a time before being released into a conducting state. Thus, an electron will drift as a free entity between traps of molecular aggregates, which are relatively immobile.

Mobilities of electrons in hydrocarbon mixtures seem to substantiate the picture of collective trapping. The variation in $\mu_e$ observed by Minday *et al.* (1972) in neopentane–n-hexane mixtures is shown in Fig. 1.19. The log of $\mu_e$ is linear with the mole fraction $X_h$ of hexane, indicating that the mixtures form nearly perfect solutions where the electron is scattered by collective interaction with each component. Similar findings have been presented by Beck and Thomas (1972) for n-hexane–iso-octane mixtures. Minday *et al.* (1972) obtained further support for a trapping model from the temperature dependence of $\mu_e$ in the mixtures. The results obeyed the Arrhenius-type eqn (1.2) and are summarised by:

$$\mu_{mix} = \mu_{np} \exp(-X_h W/kT), \tag{1.9}$$

where $\mu_{np} = 7 \times 10^{-3}$ m$^2$ V$^{-1}$ s$^{-1}$ is the mobility in pure neopentane and $W$ is the activation energy characterizing the escape of the electron from a trap.

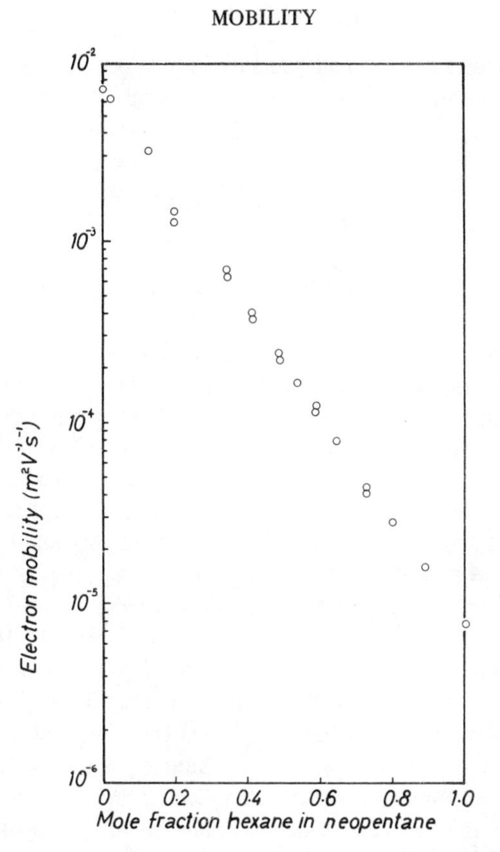


FIG. 1.19. Electron mobilities in n-hexane–neopentane mixtures (after Minday *et al.* 1972).

Values of 0.022 eV and 0.186 eV were deduced for $W$ in the pure forms of neopentane and n-hexane, respectively. It is surprising to find that in n-hexane the activation energy for mobility of electrons is somewhat greater than for slow negative ions (sub-section 1.3.1). Nevertheless, this observation would seem to reinforce the models of short-lived molecular trapping and of permanent impurity trapping for electrons in ultra-pure and 'impure' hydrocarbons, respectively. Thus, $W$ for a slow carrier is a measure of the ease with which a stable negative ion and its surrounding sheath of polarized molecules can slide through the liquid. On the other hand, $W$ for a fast carrier represents the energy necessary to release it from a molecular trap before it can move as a free electron. However, the idea of molecular traps as the purely 'fortuitous coincidence . .' envisaged by Schmidt and Allen (1970 *b*) is difficult to reconcile with the fact that an electron must interact with the liquid. It may be that the electron mobility reflects an element of self-imposed trapping which involves some molecular rearrangement. This process could relate to the anisotropy of molecular polarizability postulated by Dodelet and Freeman (1972) to explain the thermalization range of electrons in hydrocarbons.

Davis, Schmidt, and Minday (1971, 1972) proposed that eqn (1.9) be written as

$$\mu = (\mu_0/\nu\tau_0)\exp(-W/\mathrm{k}T)\,, \qquad (1.10)$$

where $\mu_0$ is the free electron mobility and $\nu$ is the frequency of trapping collisions. The residence time of an electron in a trap is expressed as $\tau = \tau_0 \exp(W/\mathrm{k}T)$ where $\tau_0$ represents the period of a molecular vibration ($\sim 10^{-13}$ s). A similar expression to eqn (1.10) was derived by Frommhold (1968) to describe electron motion in a gas with short-lived trapping. Using their extension of the Cohen–Lekner theory (sub-section 1.4.1) which accounted for inelastic scattering by polyatomic molecules, Davis *et al.* (1972) verified the magnitude of the free electron mobility in eqn (1.10) for neopentane–hexane mixtures. For hydrocarbons, in general, they concluded that $\mu_e$ should be field independent up to high fields. For neopentane, in particular, this field was estimated to be 40 MV m$^{-1}$. However, measurements by Bakale and Schmidt (1973 *b*) in Fig. 1.20 show that $\mu_e$ is neopentane follows an $E^{\frac{1}{2}}$ dependence above 0.5 MV m$^{-1}$. Nevertheless, the results in Fig. 1.20 provide extra evidence for a trapping model. In ethane the electron drift velocity increased more than linearly with field, above 8 MV m$^{-1}$. This increase is easily explained by a field-assisted contribution to the release of electrons from traps. These results represent the first direct experimental evidence for such a process in dielectric liquids. The linear increase of the drift velocity in neohexane under an applied field up to 15 MV m$^{-1}$ reflects a fortuitous balance between field-aided ejection from traps, as in ethane, and quasisaturation in velocity, as in neopentane.

Apart from mobility and pulse radiolysis studies two other experiments can provide information about the conducting states of electrons in hydrocarbons. These experiments are (1) biphotonic ionization and (2) photoelectric work-function measurements (see sub-section 1.3.4). By means of (1), Takeda, Houser, and Jarnagin (1971), and Devins and Wei (1972) have deduced estimates of $\mu_e$ in good agreement with results from time-of-flight measurements. Biphotonic ionization occurs during pulsed, or steady, light excitation of a dilute solution of a suitable dopant in a liquid. Using this technique organic solutions can be photoionized with light quanta whose energies are as little as one-half those required for gas-phase ionization. The essential processes in *bi*photonic ionization occur in *two* stages. Firstly, an excited triplet state $M^x$ of the solute molecule M is created by collision with a photon. This stage is represented by the reaction

$$M + h\nu \rightarrow M^x\,. \qquad (1.11)$$

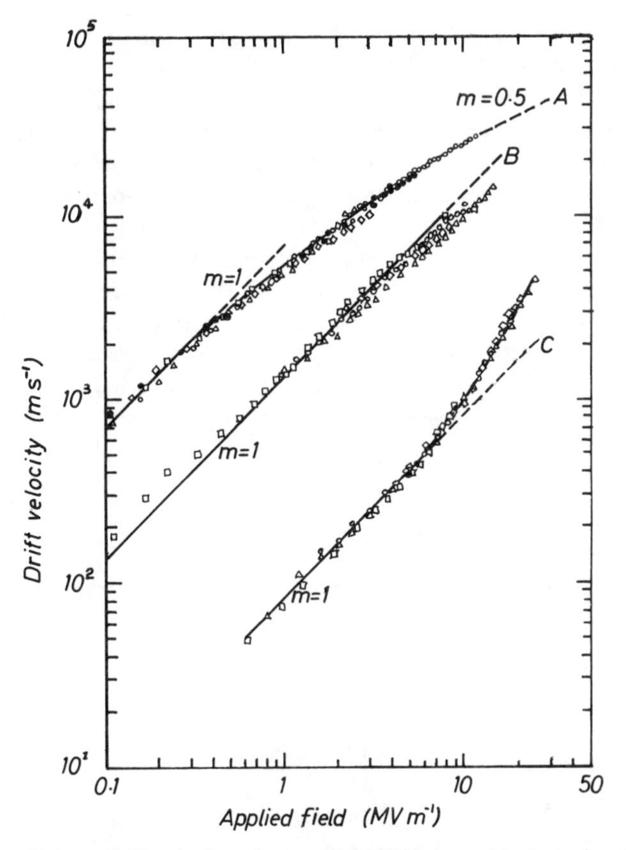


FIG. 1.20. Electron drift velocity versus applied field in several hydrocarbon liquids. Curve A, neopentane ($T = 295$ K); curve B, neohexane ($T = 295$ K); curve C, ethane ($T = 200$ K). Different symbols represent different experiments and test cells; $m$ is the slope of the dashed line (after Bakale and Schmidt 1973*b*).

Secondly, before $M^x$ can decay to the ground state, it is ionized by a second photon to create a positive ion and an electron. This reaction is shown as:

$$M^x + h\nu \rightarrow M^+ + e^-. \quad (1.12)$$

These electrons have only epithermal energies because of the small energy difference between the ionization potential of the solute molecule and the total energy of the two quanta. Tetramethyl-*p*-phenylenediamine (TMPD), at concentrations of $10^{-3}$ to $10^{-4}$ mole, is usually used as the solute molecule, although Beck and Thomas (1972) have used anthracene and pyrene in several alkanes.

Using a high-speed pulsed nitrogen laser as the light source, and high-speed electronic methods of detection, Devins and Wei (1972) have observed the transient photocurrents produced in a solution of TMPD in n-hexane. The laser supplied 1 millijoule in approximately 10 ns and the photon wavelength was 3371 Å, corresponding to a quantum of energy of 3.68 eV. Because the resolution time is so short with this equipment the behaviour of electrons can be examined within a few nanoseconds of their creation, even in normally 'impure' liquids. Thus, the rigorous purification techniques of Minday *et al.* (1971) are not necessary in these experiments. Devins and Wei (1972) related the transient photocurrents to the range of the electrons before they were trapped. As shown in Table 1.4.2 the trapping distance $\lambda$ in degassed n-hexane is, at least, a factor of ten greater than in an air-saturated sample. This observation is consistent with reduced trapping due to a smaller level of dissolved oxygen in the liquid.

Table 1.4.2.
*Trapping distances as a function of relative light intensity for n-hexane (after Devins and Wei 1972).*

| $I/I_0$ | $\lambda_{\text{air-saturated}}$ (Å) | $\lambda_{\text{degassed}}$ (Å) |
|---|---|---|
| 0.31 | – | 22 300 |
| 0.36 | – | 14 100 |
| 0.45 | 1260 | 15 200 |
| 0.57 | 1100 | 14 800 |
| 0.66 | 1420 | 8760 |
| 0.83 | 950 | 9600 |
| 1.00 | 1670 | – |

The large values for $\lambda$ in Table 1.4.2 require some comment. Trap concentrations may be approximated by taking the cube of the inverse of the mean trapping distance. Thus, $\lambda_{\text{air}}$ and $\lambda_{\text{degassed}}$ roughly correspond to impurity concentrations of 1 in $10^6$ and 1 in $10^9$, respectively. These low values are wholly unrealistic, the latter approaching the liquid purity achieved by Minday *et al.* (1971). It would appear, therefore, that for epithermal electrons in hydrocarbons, the probability of attachment is small and is not as strong as hitherto supposed. Schmidt and Allen (1970 *b*) have also noted that dissolved oxygen in hydrocarbons seem to have a capture cross-section significantly less than for other electron scavengers. The values for $\lambda$ in Table 1.4.2 can explain the failure to detect fast electrons in pre-1969 studies of charge mobilities in hydrocarbons. In degassed n-hexane the excess electrons are scavenged within approximately 20 000 Å. This distance is less than $10^{-2}$ per cent of the typical emitter-collector spacings used in mobility experiments and, therefore, the injected electrons were converted into slow negative carriers within a short distance of their source. A further point should be noted concerning the large $\lambda$ values. Holroyd, Dietrich, and Schwarz (1972) have estimated 40 Å to be the thermalization range $b$ of photoinjected electrons (energies between 0.1 and 0.5 eV) in n-hexane. By inference the electrons generated by Devins and Wei (1972) should have reached thermal energy in about 40 Å also. The large difference between $\lambda$ and $b$ has a significant meaning. It implies that a thermal electron can travel a very large distance before undergoing a collision in which there is a large energy transfer to the fluid and a large change in the directed motion of the electron. Consequently, a thermal electron must regularly lose its energy gain from the applied field to low-energy molecular collisions. Induced rotation and vibration excitations are obvious sinks for this energy drain; they may act as the main barrier to an electron reaching high energies. We shall refer to these trapping distances again in sections 2.3 and 3.7.

From the results of photoelectric work function measurements, Holroyd and Allen (1971) deduced an empirical correlation between the mobility values of Schmidt and Allen (1970 *a*) and the energy, $V_0$, needed to inject an electron from vacuum into a conducting state in a liquid (see eqn 1.7). For low mobility liquids $V_0$ was close to zero but became negative as $\mu_e$ increased, as shown in Table 1.4.3. The values for $V_0$ have been used by Fueki, Feng, and Kevan (1972) to estimate the molecular scattering cross-sections, $\sigma_l$, offered to electrons in various hydrocarbons. The results are included in Table 1.4.3 which shows that in some of the liquids $\sigma_l$ was, at least, twice that in the corresponding gas. Moreover, $\sigma_l$ increased with an increase in the number of carbon atoms in a molecule. Furthermore, for compounds with the same number of carbon atoms $\sigma_l$ decreased from the straight chain pentane, through the branched neopentane, to the cyclic form of pentane. As we shall see in sub-section 3.9.1, these observations are relevant to theories of breakdown in dielectric liquids. Takeda *et al.* (1971) and Holroyd (1972) have also demonstrated a relationship between

Table 1.4.3.
*Estimate of scattering cross-sections in liquid alkanes (after Fueki et al. 1972)*

| Liquid | $\sigma_l$ ($Å^2$) | $\sigma_{gas}$ ($Å^2$) | $V_0$ (eV) | $\mu_e$ ($m^2 V^{-1} s^{-1}$) × $10^4$ |
|---|---|---|---|---|
| n-Hexane | 59.2 | 26.5 | 0.04 | 0.09 |
| n-Pentane | 51.5 | 17.0 | –0.01 | 0.16 |
| Neopentane | 46.3 | – | –0.43 | 55 |
| Cylopentane | 43.2 | – | –0.28 | 1.1 |
| 2,2,4-Trimethylpentane | 69.8 | – | –0.18 | 7 |

$V_0$ and the threshold photon energy for ionization of TMPD in various non-polar solvents.

In theoretical attempts to explain the large variation in electron mobilities in hydrocarbons, Schiller (1972) and Kestner and Jortner (1973) have postulated a quantitative correlation between $\mu_e$ and $V_0$. However, experimental measurements of $V_0$ by Holroyd and Tauchert (1974) do not substantiate either of the theories. A novel idea was used to test the theories. Firstly, $V_0$ was measured in several neopentane–hexane mixtures. Eqn (1.9) was then used to determine $\mu_e$ for each mixture before calculating the corresponding $V_0$ predicted by each theory. The measured and calculated values of $V_0$ did not agree. Theories have been developed which can account for much of the behaviour of electrons in monatomic liquids (see sub-section 1.4.1). It is a much more formidable problem to provide a realistic interpretation of electron mobilities in polyatomic fluids but, no doubt, the problem will be solved.

### 1.5. Summary

Slow negative ions in many insulating liquids result from the capture of electrons by impurities. Oxygen appears to be the dominant impurity but other electron scavengers may also be present. In liquid helium, however, the model for a negative ion is a self-trapped electron in a cage of helium atoms. Theoretical treatments of the electron–He atom interaction have successfully described this model. The true mobilities of negative and positive ions in liquids have yet to be measured; EHD effects have produced anomalously high values in previous measurements.

Very fast quasifree electrons can exist in argon, krypton, xenon, and methane. For liquid argon the Cohen–Lekner theory, which does not contain any adjustable parameters, shows excellent agreement with experimental $\mu_e$ values. Fast electrons can also exist in ultra-pure hydrocarbons. Although these liquids exhibit similar chemical and physical properties, the electron mobilities range over three orders of magnitude. Molecular structure appears to influence $\mu_e$ in two ways. First, $\mu_e$ is observed to increase with increasing sphericity of the molecules. Secondly, the electrons seem to lose their energies in the excitation of molecular rotations. Liquid structure is also important. $\mu_e$ is determined by a series of jumps between short-lived traps, which are created by particular configurations of groups of molecules. Nanosecond detection circuits have revealed that the electron trapping distance is large in n-hexane. In degassed liquid the distance is of the order of 2 $\mu$m. Several experimental methods have been developed for the study of these fast carriers, but theoretical treatments have not yet accounted for the great range of $\mu_e$ in polyatomic fluids (see Note 9, Appendix).

In conclusion, mobility studies have provided a very deep insight into the conducting and, perforce, nonconducting states of insulating liquids.

# 2

# Conduction

## 2.1. Introduction

Conduction in dielectric liquids has been reviewed many times. Short accounts of the subject have been given by MacFadyen (1955) and by Swan (1962 *b*). More comprehensive treatments can be found in the articles by Lewis (1959), and by Sharbaugh and Watson (1962), whilst recent work on conduction was surveyed thoroughly by Zaky and Hawley (1973). Conduction, induced by the deliberate radiation of liquids, has also received considerable attention. This topic was discussed by Hummel and Schmidt (1971) and it has received voluminous coverage by Adamczewski (1965, 1969). Current flow in polar liquids was reviewed by Felici (1971 *b*). It would appear that little of interest remains to be said about conduction; a detailed survey of the subject would be mere repetition of these earlier works. Nevertheless, in this chapter, we shall devote our discussion to some of the major controversial issues concerning the natural conductivity of liquids which still require satisfactory explanations. Hopefully, by drawing on our knowledge of charge transport processes from Chapter 1, we may be able to elucidate further a few of these issues. A discussion of radiation-induced conductivity is omitted because of the extensive treatment given to it in earlier reviews.

## 2.2. Conduction at low fields

### *2.2.1. Current due to ionic impurities*

The general shape of a current–field characteristic which is usually obtained for a dielectric liquid can be roughly divided into three regions. At low fields, $< 100$ kV $m^{-1}$, the current appears to rise linearly with field, and to obey Ohm's law. At intermediate fields up to about 2 MV $m^{-1}$, there is a sub-ohmic region where the current tends towards saturation. However, the saturation region may be ill-defined or may not be present at all, as was observed by Green (1955). At high fields the current increases rapidly and, in the region of $10^8$ V $m^{-1}$, breakdown can occur. A good example of the general pattern of a current-field measurement is illustrated by the results in Fig. 2.1. The results were obtained

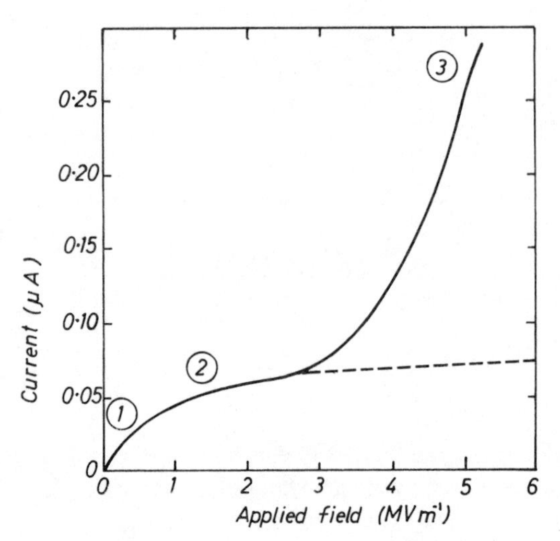


FIG. 2.1. Current-field characteristic for highly purified nitrobenzene (after Brière and Gaspard 1968).

by Brière and Gaspard (1968) for highly-purified nitrobenzene, but similarly-shaped curves have been measured for many non-polar liquids.

For a continuous current to flow, there must be a source of charge. Several agents could contribute to the current in the ohmic and sub-ohmic regimes. Amongst these are natural radiation, dissociation of ionic impurities, and solid impurity particles. The role of particles in conduction will be considered in sub-section 2.3.1. The influence of natural radiation on the current magnitude is negligibly small. Under normal conditions, air at ground level contains approximately $10^9$ ions per $m^3$ due to u.v. light, cosmic rays, etc. For a dense medium, such as a liquid, the number will be much less. Adamczewski (1969) has estimated that approximately 1700 eV is needed to create an ion pair which actually contributes to a radiation-induced current in an organic liquid. The corresponding energy for a gas is only of the order of 30 eV. Thus, the effects of natural radiation can be neglected in conduction studies on organic liquids. It is interesting to note, however, that Willis (1966) concluded that the cryogenic fluids $H_2$, $D_2$, Ar, CO, and chlorotrifluoromethane ($CClF_3$) were intrinsically perfect insulators, but that their residual, minute conductivities were produced by cosmic ray ionization.

Chemical impurities are the most likely source of charge at low and at intermediate fields, as it is impossible to remove all traces of impurities. The mechanisms of charge generation, however, are open to speculation, as is the nature of the residual contamination. The host liquid may dissociate the impurity molecules into ion pairs, or traces of highly-polar media, like water, may self-dissociate. With a small field applied to the electrode–liquid system, the dissociated negative and positive ions, which escape recombination, will drift to the electrodes and give rise to a flow of current. The ionic drift velocities and, therefore, the current will increase with increasing field, as in the ohmic region of Fig. 2.1. At higher fields the ions may be removed to the electrodes at a rate faster than their generation rate in the bulk of the liquid. The current is then determined by the generation rate, and also by the rate of ion neutralization at the electrodes. Under these conditions, regions of space charge can grow at both electrodes (*cf.* Zaky and Hawley (1973) for a discussion of space charge effects at low fields). The combined effects of bulk generation and electrode neutralization can then produce a situation where the current tends to saturation. Thus, ionic impurities can account for the conduction current at low and intermediate fields. The origin of the currents at high fields will be discussed in section 2.3.

### *2.2.2. Conduction in aromatic hydrocarbons*

There has been an extended debate in the literature concerning the mode of conduction in aromatic hydrocarbon liquids. Benzene, being the simplest of these unsaturated compounds, has received special attention. In benzene, and six methyl-substituted benzenes, Forster (1962, 1964 *a, b*, 1967) observed nearly a hundredfold increase in conductivity, compared with his results for several saturated aliphatic compounds. He interpreted the large increase on the basis of intermolecular transfer of the $\pi$-electrons associated with the benzene ring. On the other hand, Silver (1965) proposed that low-field conduction could be accounted for by the theory of Thomson and Thomson (1928) for radiation-induced conduction in gases. The essential features of this theory are (1) that the electrodes do not inject charges but only govern their rate of neutralization, and (2) there is a *thermal* equilibrium distribution of charges generated in the bulk of the liquid. Silver (1965) applied the theory to Forster's (1962) results on benzene and got good agreement, as shown in Fig. 2.2. No mechanism was postulated by Silver to provide the equilibrium distribution of charges. However, at low fields, dissociation of impurities is the only tenable process: thermal ionization of the liquid molecules is infinitesimal.

Further evidence that impurity ions constitute the major charge carriers in aromatic liquids is provided in Table 2.2.1, which contrasts reported values for the conductivity of benzene. Although a conductivity in the region of $10^{-12}$ (ohm m)$^{-1}$ is common to many of the results, some of the values differ by nearly five orders of magnitude. Experimental errors cannot be blamed for this difference! Having obtained a conductivity almost identical in magnitude to Forster's (1962) value, Pitts *et al.* (1966) abandoned conventional purification methods in favour of a bubble-cap fractionating column. Apparently, this apparatus removed more impurities, since a substantially lower conductivity of $10^{-15}$ (ohm m)$^{-1}$ was measured. Impurities, possibly released from glassware during distillation, were probably responsible for the more common conductivities of $\sim 10^{-12}$ (ohm m)$^{-1}$. It is pertinent to note that, for electron mobility studies, Minday *et al.* (1971) found great difficulty in purifying

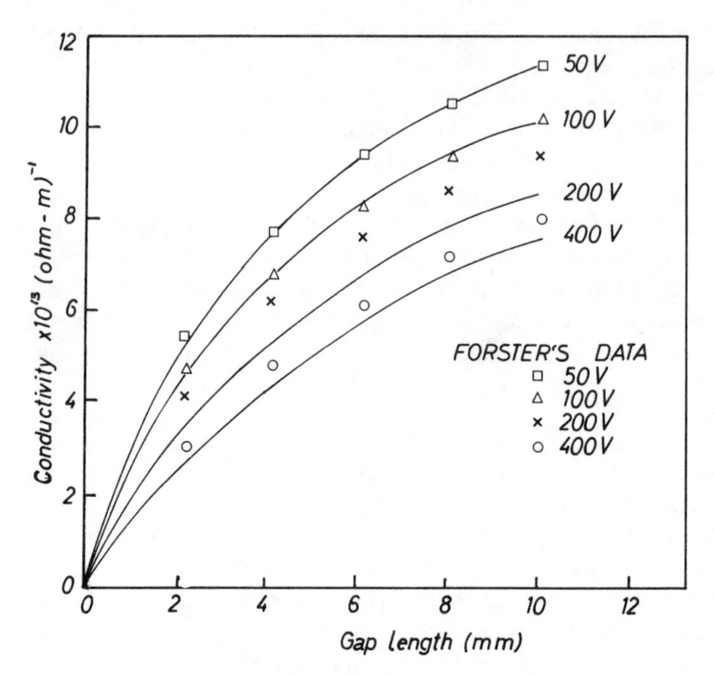


FIG. 2.2. Comparison between theoretical (solid lines) and experimental (symbols) conductivities in benzene as a function of gap length (after Silver 1965).

Table 2.2.1.
*Conductivities of benzene*

| Conductivity (ohm m)$^{-1}$ × $10^{13}$ | | Investigator | |
|---|---|---|---|
| d.c. | a.c. (1kHz) | | |
| <3 | – | Kraus and Fuoss | (1933) |
| 11 | – | Forster | (1962) |
| 7 | – | | |
| 37 | 46 | Forster | (1972) |
| 64 | – | Sano and Akamatsu | (1963) |
| 33 | – | | |
| 7 | – | Pitts, Terry, and Willetts | (1966) |
| $10^{-2} \rightarrow 2 \times 10^{-3}$ | – | | |
| 50 | – | Garben | (1974) |
| $6 \times 10^2$ | $5.4 \times 10^2$ | Vij and Scaife | (1974) |

benzene or toluene. For these liquids, at least 80 per cent of the current was ionic in character. This was ascribed by Minday *et al.* to electron-scavenging components which could not be removed by their purification techniques. Furthermore, from the results of many experiments, Guizonnier (1961, 1968) has imputed low-field conduction in liquids, including benzene, to traces of moisture, which may dissociate into ions or move, as charged microscopic droplets, between the electrodes.

If ionic conduction is the real mode of charge transport in aromatic liquids, then the current should decrease with increasing applied hydrostatic pressure because of a higher liquid viscosity (see eqn 1.3). However, Garben (1972) has reported that, at different isothermals, the current in dry benzene increased substantially with pressure up to 5 kbar. The results are shown in Fig. 2.3. A similar variation was found in water-saturated benzene (Garben 1974). These observations would appear to eliminate ionic transport in favour of the mechanism of inter-molecular hopping of electrons, originally postulated by Forster. Vij and Scaife (1974), on the other hand, have obtained results in direct contrast to those of Garben: as may be seen from Fig. 2.4, the a.c. conductivity of benzene decreased with pressure up to 3.4 kbar. Vij and Scaife (1974) have also noted that measurements of the d.c. conductivity may be affected by low shunt resistances of high-pressure electrical leads into the test

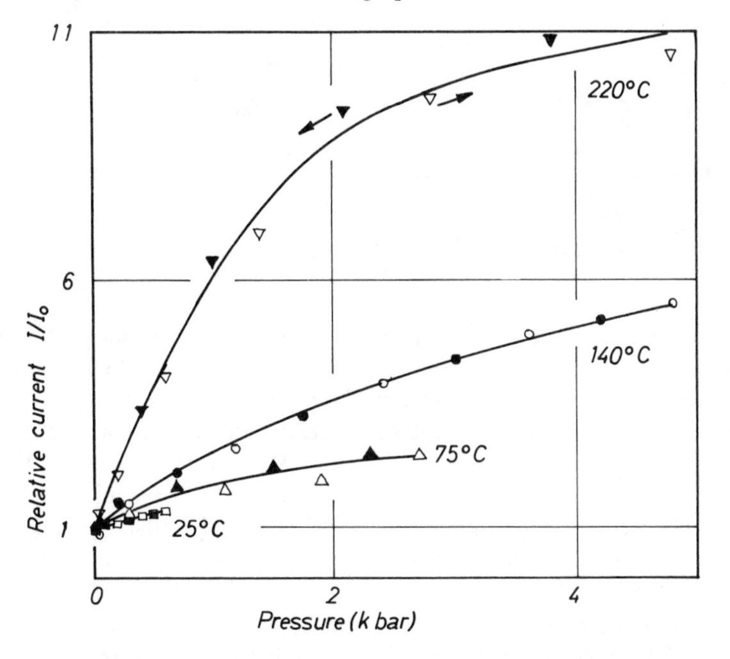


FIG. 2.3. Relative d.c. current in benzene as a function of applied hydrostatic pressure. 1 mm gap length between cylindrical brass electrodes. 800V applied at 75° C and 220°C; 12V applied at 25°C and 140°C (after Garben 1972).

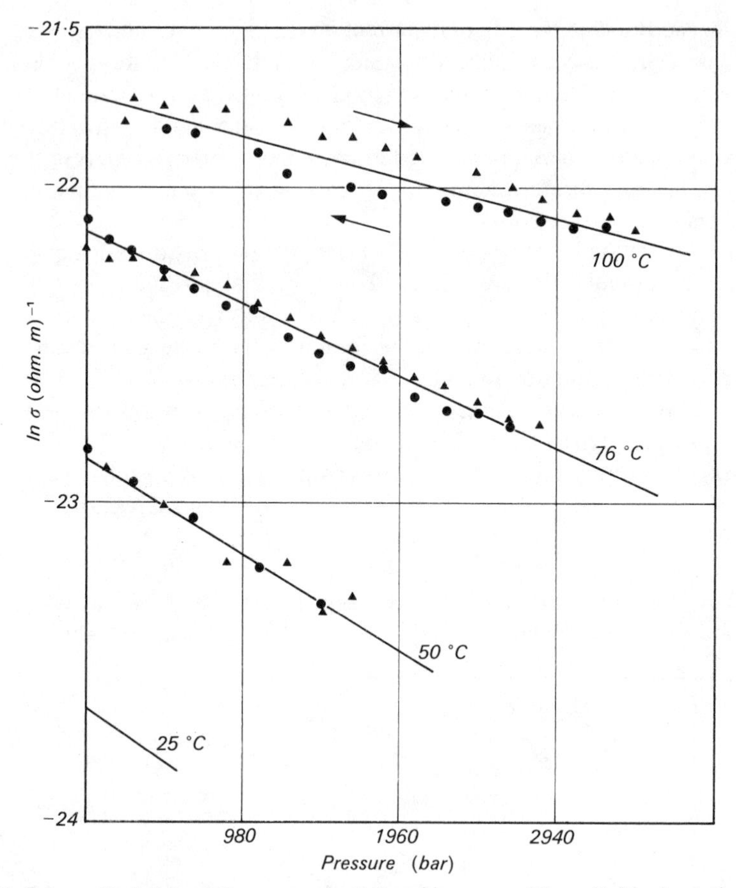


FIG. 2.4. Variation of the a.c. conductivity of benzene with applied hydrostatic pressure (after Vij and Scaife 1974).

cell. Nevertheless, their d.c. results followed a pattern similar to those in Fig. 2.4. At present, a satisfactory explanation of the conflicting results of Figs 2.3 and 2.4 is not available. The only other organic liquids which have exhibited an increased conductivity with pressure are diethyl ether, ionic solutions in diethyl ether, and castor oil above 2.5 kbar (Scaife 1974). The rarity of a pressure effect in simple hydrocarbons would seem to indicate that some extraneous factors influenced Garben's results. Contamination of the benzene by the hydraulic fluid cannot be ruled out. This possibility, coupled with the fact that the ionic dissociation of water is highly temperature-dependent (Maron and Prutton 1970), could produce a very complex electrode–liquid system. High-pressure experiments on the a.c. conductivity of benzene saturated with water may resolve the conflict between the results in Figs 2.3 and 2.4.

*2.2.3. Current due to molecular complexes*

An entirely new approach to conduction was suggested recently by Fournié (1970). His model is based on the idea of the formation of molecular complexes in liquids (cf. Rose 1967 for a reasonable account of molecular complexes). Briefly, electron transfer occurs between donor (D), and acceptor (A), molecules resulting in the formation of a molecular complex $A^-D^+$. Examples of acceptors listed by Fournié include iodine, oxygen, water, quinones, sulphur dioxide, and mineral salts. Donor molecules may comprise aromatic hydrocarbons, especially benzene and diphenyls, alkenes, and alkynes. Experimental evidence of molecular complex formation was presented by Bohon and Claussen (1951), who used u.v. spectroscopy to investigate the solubility of aromatic hydrocarbons in water. Interaction between water and the solute molecules was attributed to the labile $\pi$-electrons of the benzene ring. Dewar (1946) and Mulliken (1950) have stated that a $\pi$-complex can account for the abnormal colours and dipole moments exhibited by solutions of iodine in benzene. It should be emphasized, however, that *no* additional charge carriers arise from complex formation of the type $A^-D^+$. Hence, *no* change in electrical conductivity should take place. Fournié (1970) has suggested that charge carriers are created by field-assisted dissociation of the complex $A^-D^+$ into $A^-$ and $D^+$ ions, along the lines of Onsager's (1934) treatment of the non-ohmic behaviour of weak electrolytes.

Durand and Fournié (1970) have observed an increase in the current in a chlorobiphenyl liquid after the addition of acceptor molecules of iodine, antimony pentachloride ($SbCl_5$), or anthraquinone. The increase in current due to these additives was greater than when the liquid was saturated with oxygen or sulphur hexafluoride ($SF_6$), as shown in Fig. 2.5. The Onsager theory of dissociation adequately accounted for the results with iodine and $SbCl_5$, but it did not account for the currents due to oxygen, $SF_6$ or the anthraquinones. No explanation has been found for the behaviour of these three additives.

A model of conduction based on molecular complex formation should be particularly suited to the unsaturated aromatic liquids. Kilpatrick and Luborski (1953) have stated that increasing methyl substitution should increase the electron-donating ability of the aromatic molecule. We should expect, therefore, that conductance in the liquids benzene, toluene, 1,2-dimethylbenzene, and 1,2,4-trimethylbenzene should increase in that order. In fact, the measurements by Forster (1962, 1964 *a, b*) show that benzene had the highest conductance, whilst the substituted compounds exhibited only a slight variation. The absence of charge transfer complexes in saturated compounds, such as the alkanes (Mulliken 1950), might explain their lower conductivities relative to aromatic liquids (sub-section 2.2.2). Obviously, more experimental work is required to determine the role, if any, of molecular complexes in the conduction processes

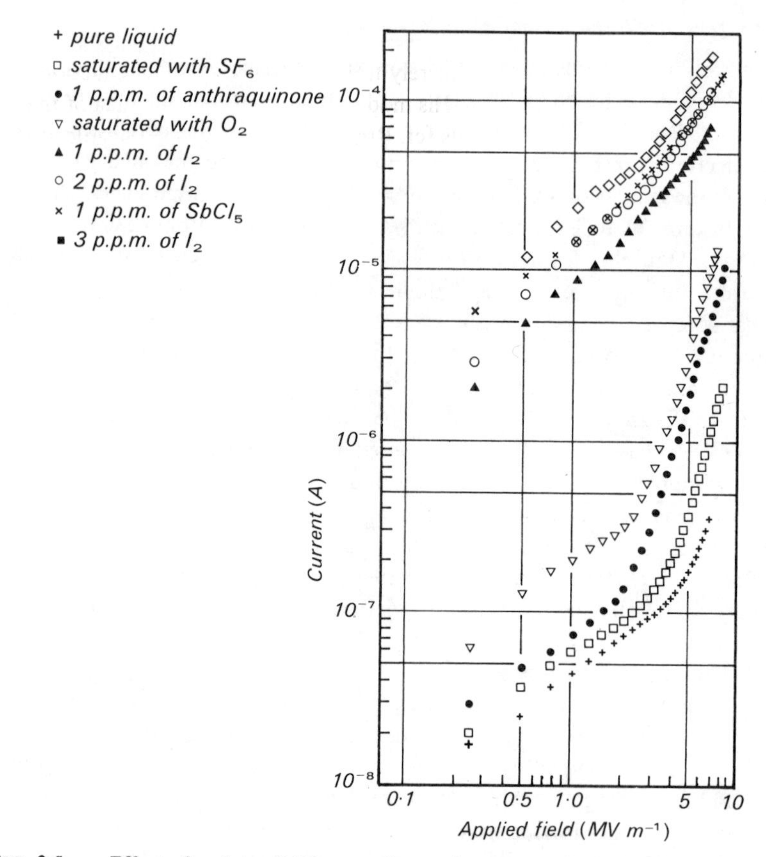


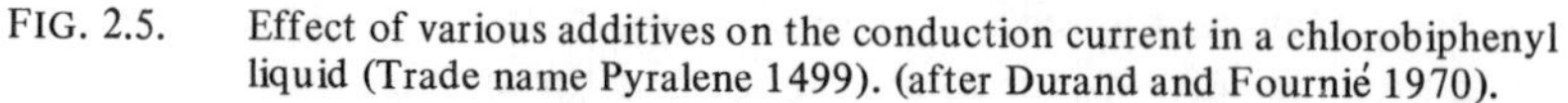
FIG. 2.5. Effect of various additives on the conduction current in a chlorobiphenyl liquid (Trade name Pyralene 1499). (after Durand and Fournié 1970).

in liquids. At present, however, we must conclude that ionic impurities are the most plausible cause of current flow at low and at intermediate fields.

## 2.3. Conduction at high fields

Interest in the study of conduction currents at high fields stems, in great part, from the belief that the mechanism of breakdown is closely linked with the process of conduction. When the field applied to an insulating liquid is increased beyond about $10^7$ V $m^{-1}$, the current tends to rise sharply for comparatively small increments of field, and at sufficiently high fields, breakdown occurs. This behaviour is typified by region 3 of Fig. 2.1. Zaky and Hawley (1973) have given a comprehensive review of the work on high-field conduction currents which has been published in the last decade. Here, we shall discuss the various suggestions for the source of these currents, as this topic remains one of the most controversial issues in this field of research. The controversy has developed in two stages. Only recently it was held that electron emission from the cathode was the current-controlling process and the major argument was concentrated on the presence (or absence) of electron multiplication by collision ionization as an extra contributory factor. Now, several workers contend that the total current is transported by a small number of sub-micron solid particles.

### *2.3.1. Current due to solid particles*

Most investigators have endeavoured to minimize the effect of solid impurities by including filters in the liquid stream at a point directly before it flowed into the test cell. Sintered glass filters were used to remove particles down to 1 $\mu$m size and, recently, membrane filters with pore sizes down to 10 nm have been employed. Repeated distillation and flushing of the cell were assumed to remove all particles of a significant size from the liquid and the electrodes. Consequently, conduction measurements at high fields were explained on the basis of mechanisms which did not include effects due to particles. The situation changed when Krasucki (1968) presented a theoretical model of particle motion in a liquid which accounted for the magnitude and the shape of the current–field characteristics.

Solid impurities can be carried into the interelectrode gap from numerous sources. The amount of contamination which could be released from the surfaces of a conventional test cell and electrode–liquid system was vividly illustrated by Krasucki (1972). Measurements were made of the particle size distribution in n-hexane after ultrasonic agitation of the test cell. Then, the sample of liquid was passed through a filter of 0.4 $\mu$m porosity, and the total number of particles removed was counted, using an image-analysis computer system. Curve (a) of Fig. 2.6 shows that over 1.1 million particles with diameters greater than 1 $\mu$m were removed in a first cleansing of the cell. Curves (b) to (f) show the number removed after repeated ultrasonic cleansing. At first glance, these enormous numbers of particles are rather alarming. It must be remembered, however, that the strong cavitation effects associated with ultrasonics can erode, quite easily, surfaces such as glass, perspex, and aluminium, which are typical of the materials used in test cells. Moreover, the usual purification techniques would not produce such deleterious effects in a cell, and it is virtually certain that previous conduction (or breakdown; see section 3.8) measurements have not involved such large concentrations of particles. Nevertheless, as suggested by Krasucki (1972), some of the particles detached from the cell walls and from the electrodes, due to ultrasonics, may be attracted into the gap under the influence of the high fields used during conduction (or breakdown) measurements. Sletten and Lewis (1963) have reported that microscopic examination of n-hexane, during high-field experiments, revealed particles of the order of 1 $\mu$m size in a 50 $\mu$m gap. The main source of particles was fresh electrodes introduced into the cell, and even vigorous flushing failed to dislodge them. From the foregoing

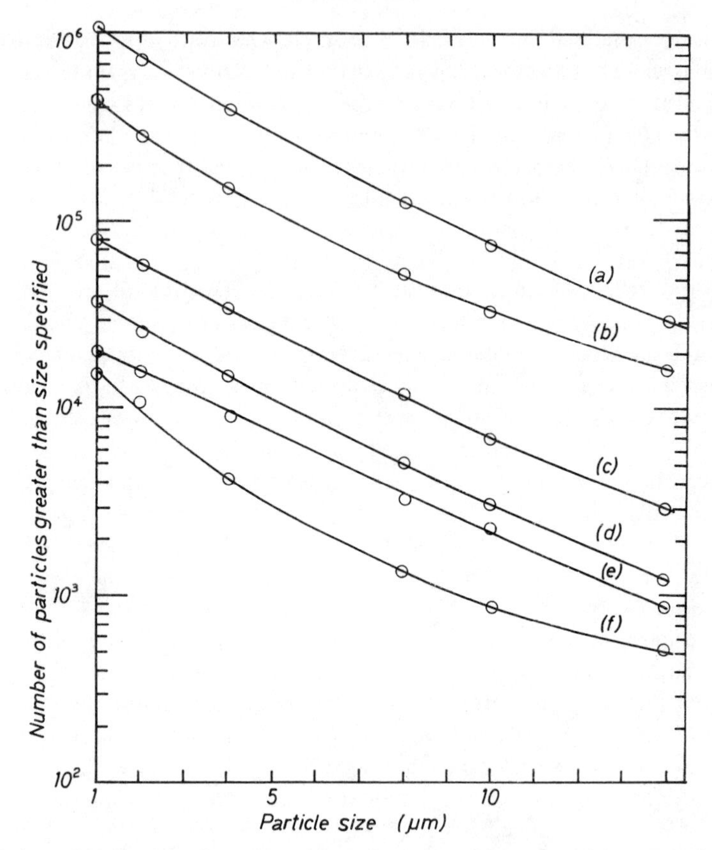


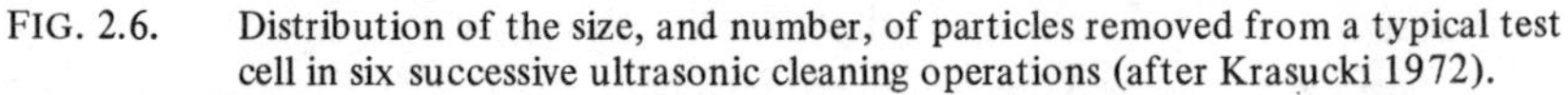

FIG. 2.6. Distribution of the size, and number, of particles removed from a typical test cell in six successive ultrasonic cleaning operations (after Krasucki 1972).

presentation we must conclude that, even before a measurement is taken, the high-field region of the gap may contain solid particles. We shall now examine how they may contribute to the current.

Krasucki (1968) considered the motion of a spherical particle of radius $R$ in a liquid with absolute permittivity $\epsilon$, between two plane electrodes subjected to an applied field $E$. In his model he assumed (*i*) that the sphere was conducting, (*ii*) it instantaneously charged to a value $Q$ on contact with the anode and (*iii*), due to the electrostatic force $QE$, it was attracted to the cathode, where it was instantaneously neutralized and acquired a charge $-Q$. The particle was then attracted back to the anode and this oscillatory motion between the electrodes continued ad infinitum. The value for $Q$ was given by

$$Q = (\tfrac{2}{3})\, \pi^3\, \epsilon R^2 E\,, \tag{2.1}$$

which was derived from Maxwell's (1892) treatment of the charge acquired by a sphere in contact with a perfect plane. The average current due to particle motion between the electrodes was expressed as $I = Q/\tau$ where $\tau$ was the time for a particle, having attained a steady-state velocity $V$, to cross the gap. The quantities $\tau$ and $V$ were computed using iterative techniques to solve the equations of motion for a particle and, with the aid of eqn (2.1), Krasucki (1968) calculated the dependence of current on the applied field. In Fig. 2.7, calculated current–field characteristics for n-hexane are compared with the experimental measurements by House (1957). It is seen that there is good agreement between theory and experiment if it is assumed that the measured current was carried by a single particle, or 646 particles, with radii of 188 and 1.88 nm, respectively. Krasucki (1968) concluded that previous data on conduction was governed by the motion of particles.

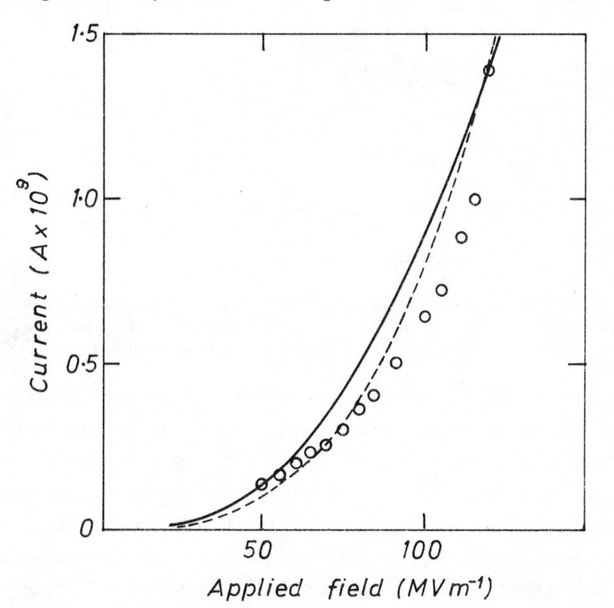


FIG. 2.7. Calculated current–field characteristics for n-hexane compared with measurements (o) by House (1957). —— Current for one particle of 188 nm radius; --- Current for 646 particles of 1.88 nm radius (after Krasucki 1968).

Birlasekaran and Darveniza (1972 *a*) examined the motion of an 800 $\mu$m diameter metallic sphere in transformer oil. Contrary to the predictions of eqn. (2.1), their results (Fig. 2.8) indicated that the charge acquired by the particle was not a linear function of the applied field. This observation conflicts with the findings of Krasucki (1968) and Dakin and Hughes (1968), but confirms those of Felsenthal and Vonnegut (1967). However, the cause of these discrepancies is unknown. On the other hand, Birlasekaran and Darveniza (1972 *a*) observed

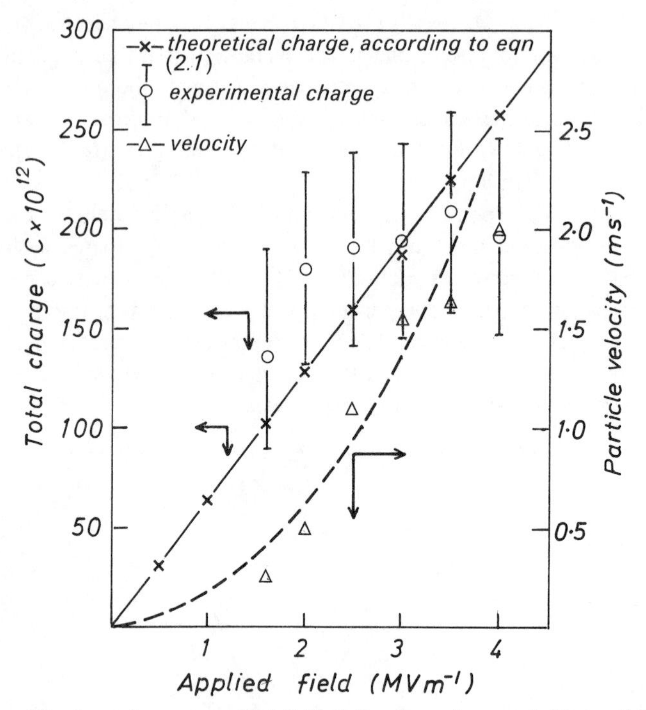


FIG. 2.8. The dependence on applied field of the charge transported by, and the velocity of, an 800 $\mu$m diameter metallic sphere in transformer oil. 5 mm gap length. (Data abstracted from Table 1 of Birlasekaran and Darveniza 1972*a*).

that the particle velocity was roughly proportional to the square of the field, in agreement with Krasucki's theory. These authors also detected pulses of sound and light at each electrode–sphere contact, with the charge being transferred to the electrode by means of a minute spark. It is interesting to note that Gzowski, Wlodarski, Hesketh, and Lewis (1966) explained electroluminescence in liquids by partial breakdowns caused by the arrival of charged dust particles at a cathode. In a theoretical paper, Birlasekaran and Darveniza (1972 *b*) applied the particle model of conduction to previous results of high-field experiments. Table 2.3.1 shows the number and size of particles required to give the best fit between calculation and experiment. The results are very reasonable and, certainly, the size of particles in Table 2.3.1 would not have been filtered out from the liquid. However, as in Fig. 2.7, the curve-fitting involved two independent parameters of size and number, and the agreement in Table 2.3.1 does not prove that particles were entirely responsible for the observed currents. In unpublished work in the author's laboratory, Krasucki's calculations were repeated for a particle with a radius of 188 nm in liquid argon, rather than in n-hexane. The current–field curve was almost identical with the corresponding curve in Fig. 2.7. Now, under experimental conditions, a test cell with liquid

Table 2.3.1.

*Number and size of carbon particles to give the best fit between calculated and experimental current–field curves (after Birlasekaran and Darveniza 1972b).*

| Liquid | Investigator | Filter pore size (μm) | Particles postulated Number | Radius (μm) |
|---|---|---|---|---|
| n-hexane at a 50 μm gap | Goodwin and MacFadyen (1953) | – | 33 or 2 | 0.12 or 0.3 |
| | House (1957) | 5 to 10 | 11 | 0.04 |
| | Sletten and Lewis (1963) | 1 | 3 | 0.04 |
| | Zaky *et al.* (1963) | 1 | 6 | 0.06 |
| Transformer oil at a 1000 μm gap | Zein El-Dine *et al.* (1965) | 1 | 460 | 0.5 |
| | Sugita *et al.* (1960) | – | 153 | 0.5 |

argon probably contains more solid impurities than one with a room-temperature liquid, where methods of filtration and flushing can be used to remove particles. However, Gallagher (1968) could not measure any *steady* current $> 10^{-14}$ A in argon for applied fields up to breakdown. Either particles of any size were practically non-existent, which is most unlikely, or charge transport by particles is insignificant in argon, and probably, in other low-temperature liquids.

Mirza, Smith, and Calderwood (1970) and Thomas (1972) showed that removal of particles by filtering reduced the conduction current. As shown in Fig. 2.9, Mirza *et al.* found that there was a twenty-fold decrease in current when particles less than 100 nm were removed from n-hexane but, surprisingly,

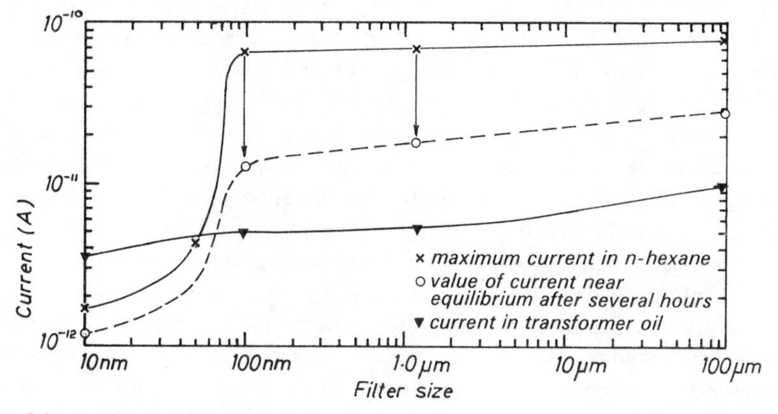


FIG. 2.9. Effect of filtering n-hexane and transformer oil on conduction current (after Mirza *et al.* 1970).

the decrease in transformer oil was very small. The results of Fig. 2.9 also indicate, however, a negligible drop in current in n-hexane or transformer oil, despite a reduction in filter porosity of $10^3$ and $10^4$, respectively. If conduction was due to charge-carrying particles, a much greater reduction in current would be expected. Unfortunately, this experiment was carried out at an applied field of only 0.5 MV $m^{-1}$ : similar work at high fields could be much more informative. Nevertheless, the curves in Fig. 2.9 are relevant to another aspect of conduction. The difference between the maximum and 'equilibrium' currents in n-hexane was interpreted by Mirza *et al.* as a gradual precipitation of particles from the interelectrode region to other parts of the test cell. Thus, particles may be responsible for the slow decay of current with time, which is generally observed in conduction measurements, but which has been attributed to a polarization process involving space charges at the electrodes (cf. Zaky and Hawley 1973). The hypothesis that particles may account for the current decay received added support from the work by Matuszewski, Terlecki, and Sulocki (1972). These authors reported that high-speed rotation of a cell containing cyclohexane greatly reduced the decay of current, but that the usual decay was restored after shaking the liquid. The natural deduction is that centrifuging the liquid caused particles to be ejected from the interelectrode gap.

Rather than deliberately introduce large metallic particles into a liquid, Rhodes and Brignell (1971, 1972) have examined charge transport in a test cell where individual 'natural' particles could be observed. The small degree of contamination was achieved, apparently, after the cell was used for several months with fresh samples of filtered n-hexane. In marked contrast to the idealized conducting particles assumed in Krasucki's theory, the results of these authors suggested that the residual particles were moderately insulating: their transit time across a 100 $\mu$m gap was approximately 10 $\mu$s, whilst their 'dwell' time on either electrode was greater than 100 ms. Somewhat similar observations were reported by Birlasekaran and Darveniza (1972 *a*) for a glass sphere in transformer oil. The long 'dwell' times indicate a very inefficient charge exchange process, which is probably caused by insulating layers on the electrodes. Two observations of Rhodes and Brignell (1972) are at variance with the findings of Birlasekaran and Darveniza (1972 *a*) for a metallic sphere. Firstly, the pulse of charge transferred to an electrode by a 'natural' particle was constant and only of the order or several femtocoulombs. Secondly, the particle velocity was proportional to the applied field between 20 and 70 MV $m^{-1}$. In addition to the small pulses ($\sim$1fC) associated with particle motion, Rhodes and Brignell observed large pulses ($\sim$1pC) at fields above about 40 MV $m^{-1}$, which sometimes culminated in a breakdown of the liquid. There appeared to be no positive correlation between the large pulses and particle motion, but particles were ejected from the gap during the appearance of these pulses. The analysis of pulse heights was combined with conduction measurements made at the same time. The total d.c. current was in good agreement with that measured by House (1957), but the amount carried by pulses did not exceed 1 per cent of this total current, as shown in Fig. 2.10. Many authors have discerned two types of pulse activity during conduction measurements. With the aid of several elegant techniques, Tropper and his colleagues have made a systematic study of the current fluctuations in transformer oil (cf. Zaky and Hawley 1973 for a full account of these measurements). Megahed and Tropper (1971) have related the large pulses to particle movement, whereas the small pulses were attributed to ionization of microscopic gas bubbles by electrons from the cathode. This observation was effectively supported by Nelson and McGrath (1972), who found that the statistical nature of the pulses was altered by induced liquid motion, which may have created cavities at the electrodes. However, it must be stated here that great care is needed in any comparison of the results of these investigations: each worker has used a different conditioning procedure (sub-section 2.3.2) which could alter results in a manner which is largely unpredictable. In fact, Nosseir (1973) has detected a change in the distribution of pulse heights with the duration of the conditioning period.

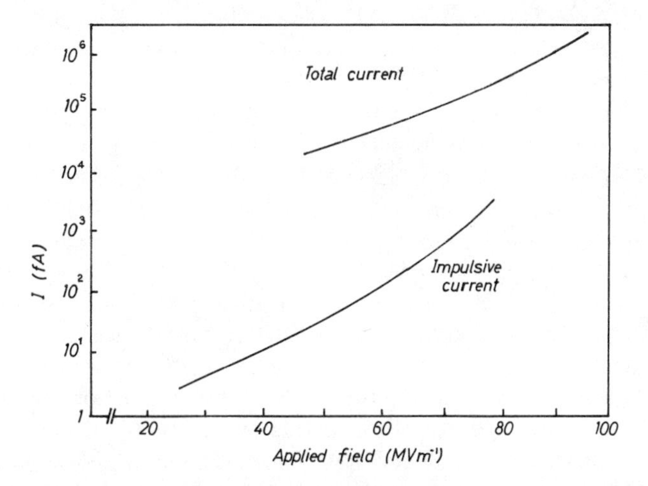


FIG. 2.10. The dependence on applied field of the total conduction current, and of the amount of current due to pulses (after Rhodes and Brignell 1972).

Felsenthal and Vonnegut (1967) have also presented results which do not substantiate a model of conduction based solely on charge transport by solid particles. These authors measured currents due to conducting particles in suspension in a commercial fluorinated liquid, marketed under the trade name of Freon 113. A tenfold change in particle radius, $R$, from 200 to 2000 $\mu$m did not change their velocities, which were only dependent on the applied voltage. These observations prompted Felsenthal and Vonnegut to suggest that the effective charge, acted on by the driving electrostatic force, was not given by eqn (2.1) but remained constant at a value $Q'$ as the field was changed, and also that the ratio $Q'/R$ was constant. It is worth noting that the results in Fig. 2.8 also indicate a constant charge, if the initial measurement is neglected. Under conditions of constant $Q'$ and with the form of Stokes's law in eqn (1.3), it follows that the particle velocity was proportional to the applied field, in agreement with their experiments. Moreover, as $Q'/R$ was apparently constant, these authors defined a maximum effective potential, $V_0 = Q'/4\pi\epsilon R$, for a spherical particle in a liquid. $V_0$ was 16 V in Freon 113. In a few experiments on n-hexane and transformer oil, however, a value for $V_0$ was measured which was dependent on particle size. No explanation was found for the different behaviour in these hydrocarbon liquids. The hypothesis of a maximum potential per particle was supported by Rhodes and Brignell (1972), who measured 12V for $V_0$ in n-hexane. Using the expression for $V_0$ with $Q' = 1$fC (small pulses) the associated particle has a diameter of 0.8 $\mu$m. This is exactly the porosity of filter used by Rhodes and Brignell, but the coincidence must be fortuitous in view of the empirical origin of $V_0$. To transport $Q' = 1$pC (large pulses), with $V_0 = 12$ V, a particle of 800 $\mu$m diameter would be required, which is clearly absurd for the gaps between 100 and 200 $\mu$m used by Rhodes and Brignell (1972). It would seem that particles were definitely not responsible for the large current pulses in their experiments.

An appraisal of the evidence presented in this sub-section suggests that mechanisms of conduction other than charge transport by macroscopic particles are required to account for the major component of current in insulating liquids. The crucial results are those in Fig. 2.10, in which are compared total current, and 'pulse' current, measurements taken at the same time. The assumption of instantaneous charge transfer between a particle and an electrode is not justified, and it would seem that the charge acquired by a particle in a liquid–electrode system cannot be described by classical electrostatics. Nevertheless, it is obvious that a small contribution to the total current can be credited to particles. The current–field characteristic for an electrode–liquid system free from physical contamination has yet to be measured. Hopefully, methods of purification which use centrifuging and membrane filters will yield a system where only electrons and ions are present.

### *2.3.2. Current due to the electrodes*

If we accept that a substantial amount of the total conduction current is not associated with charged solid particles as carriers, then the current must originate elsewhere. Electron emission from the cathode has been the most popular explanation for the source of current at high fields. The emission was assumed to follow a Schottky (1914) field enhanced thermionic process or a Fowler and Nordheim (1928) cold field tunnelling process. Briefly, Schottky showed that the potential barrier to electron emission from a metal surface into vacuum is lowered by an applied field. Consequently, the effective work function of a cathode is reduced and an enhanced thermionic current can flow. Fowler and

Nordheim (1928) showed that if the applied field is high enough the work function is decreased still more and the potential barrier is made so narrow that quantum-mechanical tunnelling occurs. A full account of these processes has been given by Lewis (1959) and they have also been considered by Sharbaugh and Watson (1962).

The lowering of the barrier height, by an amount $\Delta\phi$, due to the Schottky effect can be expressed as

$$\Delta\phi = (\tfrac{1}{2}) \left[\frac{eE}{\pi\epsilon_0\epsilon_r}\right]^{\frac{1}{2}} \text{V}, \qquad (2.2)$$

where $\phi$ is the metal work function, $e$ is the electronic charge, $E$ is the applied field, $\epsilon_0$ is the permittivity of free space, and $\epsilon_r$ is the relative permittivity of the medium in which the metal is immersed. Using eqn (2.2) with $\epsilon_r = 1.87$ for n-hexane, and fields of $10^7$ V m$^{-1}$ or $10^8$ V m$^{-1}$, corresponding to the range of electric stresses of interest in high-field measurements, we obtain reductions in $\phi$ of 0.088V and 0.28V, respectively. These are fairly significant reductions, particularly in view of the exponential dependence of emission on work function. The theoretical value of the field required to induce tunnelling into vacuum is of the order of $10^9$ V m$^{-1}$, although Llewellyn Jones (1953) has stated that tunnelling can occur in experiments on gas discharges for mean fields of $10^6$ V m$^{-1}$. This low value has been attributed to localized enhancement of the field by the combined effects of surface irregularities, oxides and positive ion layers on the cathode. Field emission into liquids has received considerable attention recently. It was investigated by Coe, Hughes, and Secker (1966), Hughes (1970) and Aplin and Secker (1972) in 'impure' n-hexane, by Coelho and Sibillot (1969, 1970) and Sibillot and Coelho (1972, 1974) in liquid nitrogen, by Halpern and Gomer (1969 *a*) in several cryogenic fluids, and by Schmidt and Schnabel (1971 *a, b*) and Schnabel and Schmidt (1973) in highly purified styrene, benzene, and $\alpha$-methylstyrene. For the pure hydrocarbons and liquefied gases, all the results indicated that the onset of a tunnelling current required fields of the order of $10^{10}$ V m$^{-1}$. For 'impure' n-hexane, the process of charge injection is not understood, but Hughes (1970) has observed that injection is apparently facilitated by the presence of water at a concentration as low as 100 p.p.m. Incidentally, these high fields were attained with an electrode geometry consisting of an etched tip or sharp edge (razor blade) with radius $<1000$ Å, opposite a flat electrode; the breakdown strength of these liquids with conventional sphere–sphere geometries is only in the region of $10^8$ V m$^{-1}$ (Chapter 3).

From the preceding arguments it would appear that Schottky emission is the more likely mechanism in liquids at room temperature but, to date, the experimental evidence in support of this is inconclusive. High-field conduction measurements show little dependence on cathode work function. In fact, it was necessary to postulate absurdly low values for the emission parameters, together with considerable field enhancement at the cathode in order to achieve any agreement between theory and experiment. This is not surprising, however. The cathode–liquid interface that was present in conduction experiments probably bore little resemblance to the idealized metal–vacuum system considered by Schottky or by Fowler and Nordheim. The electrode surfaces were microscopically very rough and were contaminated with oxides, polishing materials and adsorbed gases (section 3.4). In this state a cathode will not exhibit properties of the bulk metal and a clear dependence on work function will not be distinguished in current measurements. Also, the liquids probably contained dissolved impurities, which could alter conditions at the cathode–liquid interface. Moreover, the use of Schottky or Fowler–Nordheim laws to describe emission into liquids implies that the fluid can be considered as a structureless dielectric, where its only effect is to lower the vacuum work function of a cathode by the factor $1/\epsilon_r$ (see eqn 2.2). However, we have seen in Table 1.4.3 (sub-section 1.4.2) that an empirical correlation exists between electron mobilities in purified hydrocarbons and the effective work function of metals immersed in them. Since $\mu_e$ is influenced by liquid structure, we may assume that the structure also affects the barrier to emission. Obviously, electron emission into liquids is dependent on the nature of the metal–liquid interface and cannot be described simply in a manner similar to emission into a vacuum. Nevertheless, from the results of many workers, there is some evidence of the influence of electrode material on conduction current. Fig. 2.11 shows the d.c. characteristics obtained by Zaky, Tropper, and House (1963), using electrodes of copper and stainless steel in n-hexane. The curves follow the general shape of Fig. 2.1, with a quasisaturation zone preceding a rapid increase of current. Similarly shaped curves were obtained using mixed electrode pairs of nickel–copper and aluminium–chromium, but the extent of the quasi-saturation zone changed with each pair. Zaky *et al.* observed that there was no difference in the current if a copper electrode, which was previously oxidized for 4 hours, was used as the cathode or the anode, as shown in Fig. 2.11. Using stainless steel, however, this procedure produced a marked change in the shape of the current–field curve. With an oxidized anode the current increased rapidly for fields beyond about 25 MV $m^{-1}$. As a cathode, this electrode caused little change in the current up to 80 MV $m^{-1}$, before breakdown occurred suddenly at 88 MV $m^{-1}$. Zaky *et al.* (1963) attributed the different turning points in Fig. 2.11 to differences at the cathode, and the rising currents to field emission from the cathode. Any role the anode might play in controlling the current was neglected. We must remember, however, that for a continuous current in the external measuring circuit, negative charge must be donated at the cathode and accepted at the anode. If the cathode–liquid interface can influence emission, it is reasonable to expect that the anode–liquid interface is important for extraction. It is proposed here that the measurements by Zaky *et al.* (1963),

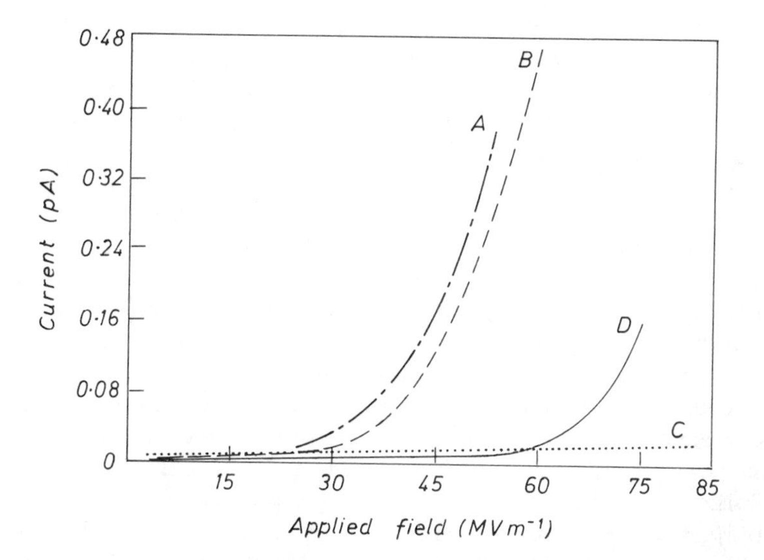


FIG. 2.11. Current-field characteristics for n-hexane at a gap length of 75 $\mu$m. Electrodes: A, stainless steel of 'zero' oxidation; B, stainless steel, with the anode oxidized for 48 hours; C, stainless steel, with the cathode oxidized for 48 hours; D, copper, with the anode or the cathode oxidized for 4 hours, opposite a cathode or anode, respectively, or 'zero' oxidation (after Zaky *et al.* 1963).

and probably most other conduction experiments, were influenced by the anode.

At this stage, it is appropriate to consider how the anode might affect conduction currents. Our discussion will be based on the mechanism postulated first by Swan and Lewis (1961) to account for the unusual control of the anode on the breakdown of liquid argon (section 3.5). Two experimental conditions are necessary in order for an anode effect to exist: (*i*) the liquid should contain an impurity which can trap electrons, emitted from the cathode, to form stable negative ions, and (*ii*) the anode should be covered with an insulating layer which can block the neutralization of these negative ions. A blocking anode can cause a negative ion space charge to build up, which limits the anode current. As a result, the field in the interelectrode region will become non-uniform, the cathode field being depressed and that at the anode enhanced. Ultimately, under conditions of direct voltage, the field in the anode region may become high enough to produce positive ions, via electron collision ionization of the liquid molecules. These positive ions, on moving to the cathode, enhance the field there to increase electron emission. This type of feedback could have a profound influence on the field distribution and lead to a rapid increase of current with applied field, as in Fig. 2.11. The proposed anode mechanism is open to the strong criticism that there is little evidence of ionization by electron collision in liquids (sub-section 2.3.4). However, if the field due to space charge at the anode is high enough, field ionization of the liquid molecules may occur. Either way, positive ions are produced so that, overall, the anode process remains unaltered. Halpern and Gomer (1969 *b*) and Schnabel and Schmidt (1973) have presented evidence of field ionization in cryogenic and hydrocarbon liquids, respectively, but the process has not been studied in detail.

Condition (*i*) above was fulfilled in all experiments on conduction. Although measurements, particularly those by Morant (1960, 1972), have been made on hydrocarbons with a high degree of purity, all the liquids may still be classed as 'impure' (sub-section 1.3.1). Condition (*ii*) was satisfied to a varying degree as different testing procedures and methods of electrode preparation were used. Most workers, naturally, desired to measure stable, and reproducible, currents. Stability was difficult to achieve because of violent fluctuations in the current (sub-section 2.3.1) but House (1957), Zaky *et al.* (1963) and others overcame the problem only after long and careful conditioning of the electrode–liquid system under gradually increasing applied field (cf. Zaky and Hawley 1973 for a full discussion of conditioning). The fact that the conditioning process itself influenced results was apparently ignored, however. Now, it is a long-established fact that wax-like polymer substances are formed on the anode during high-field conduction measurements (MacFadyen 1955). Recently, after field emission experiments on n-hexane, Aplin and Secker (1972) have observed a waxy insulating layer on the tip electrode, when it was used as an anode. This layer could only be completely removed by heating the tip to 500 °C in an oxygen atmosphere. Thus, we must conclude that in high-field measurements, especially after conditioning, charge extraction occurred at a blocking anode. Therefore, rapidly-rising curves such as those measured by Zaky *et al.* (1963) in Fig. 2.11, and by many other experimenters, may reflect a combined anode and cathode control on the current. It is significant that Zein El-Dine, Zaky, Hawley, and Cullingford (1964, 1965) observed an anode influence on conduction and space charge in transformer oil, using electrodes coated with a thin film of thermoplastic. Also, the mechanism of anode control on conduction current outlined already was invoked by Hesketh and Lewis (1969) for measurements in n-hexane and n-decane, and by Byatt and Secker (1968) in liquid air and liquid nitrogen. In Fig. 2.11 the almost constant current measured by Zaky *et al.* with an oxidized stainless steel cathode was attributed to an increase in its work function due to the oxide layer. This is the most plausible explanation (see section 3.5), but it merely reinforces our proposition that the rising curves in Fig. 2.11 were determined, to a large extent, by the anode. The absence of any difference in current in Fig. 2.11 when copper electrodes were used can be explained if it is assumed that even at zero oxidation the anode surface is sufficiently blocking to control the current. Zaky *et al.* (1963) used a high-speed polishing technique to prepare the electrodes and, because of the high surface temperatures involved, this process could readily produce an insulating oxide layer on copper, even of 'zero' oxidation.

If we accept that an anode effect can exist in conduction measurements then several inconsistencies in the literature can be explained in a qualitative manner. Firstly, the contradictory observations of different workers is only to be expected because of variations in conditioning procedure, electrode materials, impurity in the liquid etc. The shape of the current–field characteristic is determined by the combined action of the cathode and the anode, rather than the former only. Secondly, the lack of agreement with Schottky or Fowler–Nordheim laws is not surprising if emission is partly regulated by processes occurring at the anode. Thirdly, it is now possible to explain why currents in n-hexane measured under pulse voltages of a few $\mu$s duration by Watson and Sharbaugh (1960) were orders of magnitude higher than those with direct voltage under comparable conditions (sub-section 2.3.4). In order to develop, an anode effect requires a time of the same order as the sum of the transit times of a significant number of negative and positive ions. Unfortunately, we cannot estimate this time with any great accuracy. However, on the basis of mobility values (sub-section 1.3.1) it was at least 50 $\mu$seconds in the experiments of Watson and Sharbaugh and may have been much longer. Their cathode field, therefore, was not depressed by a negative space charge at the anode but, rather, the full applied field, probably enhanced by surface irregularities, was available at the cathode. Consequently, intense electron emission was observed by Watson and Sharbaugh (1960). In section 3.5 we shall see a similar pattern with respect to dielectric breakdown: the anode can determine the strength of liquids under direct voltage but does not alter it when $\mu$s pulse voltages are used.

### *2.3.3. Effect of gaseous impurities*

Gaseous impurities constitute another experimental variable which may have caused inconsistencies in the results of different investigators. Relatively little work has been published about this topic in simple dielectric liquids, whereas a detailed study has been made of the influence of dissolved impurities on conduction–current pulses in mineral oils (cf. Zaky and Hawley 1973).

Sletten (1959) was the first to notice that the addition of small quantities of dissolved oxygen to degassed n-hexane nearly doubled the breakdown strength, and it also enabled him to measure stable currents up to a field of 120 MV $m^{-1}$, which was just below breakdown. Later, however, Sletten and Lewis (1963) attributed the stable currents to traces of moisture rather than to oxygen, but they could not discern the mechanism of stabilization. At applied fields in the region of 100 MV $m^{-1}$, currents of a few nA were measured in n-hexane which contained gas in solution (see Fig. 2.11) whereas, in highly degassed n-hexane at about 20 MV $m^{-1}$, Kahan and Morant (1963) measured currents of a few $\mu$A. These high currents were very erratic and unreproducible, besides fluctuating between fairly distinct limits which were several orders of magnitude apart. Subsequently, Kahan and Morant (1965) used gas chromatography to analyse n-hexane from two manufacturers. They found that the grade used by them, and by many earlier investigators, contained up to 20 per cent of hexane isomers and other impurities, whilst the other grade was 99.99 per cent n-hexane. When the isomers were removed, Kahan and Morant (1965) observed greatly reduced current fluctuations but the processes involved are not understood (Morant 1972). Although there was also a slight improvement in the reproducibility of current measurements, Fig. 2.12 shows that considerable variation still persisted for samples of highly degassed n-hexane. These measurements were made by Morant (1972) under experimental conditions of purity etc., which were as near as possible identical. The spread in results makes it clear that complete control over all the experimental variables has not yet been attained.

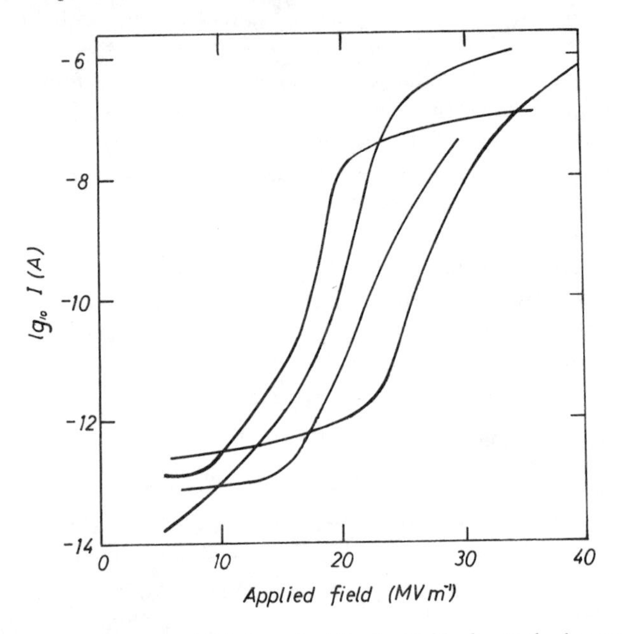


FIG. 2.12. Typical current–field characteristics for highly degassed n-hexane. Curves represent separate measurements taken in as near as possible 'identical' electrode-liquid systems (after Morant 1972).

Since small currents were measured in liquids with gas in solution, the large currents in Fig. 2.12 can be associated with a lack of gas, especially oxygen, in the degassed liquid. However, the mechanisms involved are open to speculation but, in view of the effect of oxygen on electron mobilities (section 1.3), an interpretation based on a scarcity of negative ions seems unavoidable. A shortage of negative ions precludes an anode control on the current (sub-section 2.3.2). Thus, our explanation for the large currents under pulse voltages may be substantially correct for the case of d.c. conduction in highly degassed n-hexane.

It is obvious from the preceding discussion that the effects of impurities have certainly produced many discrepancies between the observations of different investigators. If we are to gain a proper understanding of the fundamental mechanisms of conduction, it would appear that future work should be carried out along two main lines: on the one hand, ultra-pure hydrocarbons must be investigated and, on the other hand, the effects of controlled doping with additives should be studied in detail. Each experiment should also include a chromatographic analysis of the liquid in order to define precisely the concentration of chemical impurities. The results in this sub-section indicate that we have not succeeded in fully controlling all of the experimental variables. Until this stage is reached, attempts to develop a general theory of conduction in liquids are hardly justified.

*2.3.4. Current due to collision ionization*

There has been much argument and speculation as to whether electron multiplication by collision ionization contributes to the current in high-field conduction (and breakdown) measurements. As early as 1934 Nikuradse suggested that the current, at constant applied field, could be expressed as

$$I = I_c \exp \alpha d, \tag{2.3}$$

where $d$ is the gap length and $\alpha$ is a constant for the liquid and the particular value of applied field. Eqn (2.3) follows the theory of ionization by electron collisions put forward by Townsend (1910) to explain the growth of currents in gases: a current $I_c$ of electrons leaves the cathode and each electron generates $\alpha$ new electrons (and positive ions) per unit distance in a direction parallel to the applied field. If a Townsend $\alpha$-process, as collision ionization is called, exists in liquids, then, according to eqn (2.3), the larger the gap length the higher is the conduction current. In other words, the current should depend on the volume of liquid between the electrodes.

Ionization requires energies in the range of 10 eV for hydrocarbon liquids. Most of the evidence for such a process in liquids is indirect. Gap-dependent conduction currents measured by Goodwin and MacFadyen (1953) and by House (1957) for n-hexane, and by Hesketh and Lewis (1969) for n-decane, were attributed chiefly to collision ionization. On the other hand, Green (1955) was emphatic that his conduction studies of n-hexane did not reveal an $\alpha$-process, except perhaps above an applied field of 80 MV $m^{-1}$. Even at higher fields Green (1955) deduced that exp ($\alpha d$) in eqn (2.3) was not greater than 2. These investigations suffered from the use of sphere–sphere and crossed-wire electrode geometries which cannot be relied upon to maintain constant, and *uniform*, field conditions as the gap length is varied. An increase in current with increasing gap length can be taken as definite evidence for an $\alpha$-process only when *plane, parallel*, electrodes are used. Moreover, the results, referred to earlier, were based on d.c. measurements. As we have seen in sub-sections 2.3.2 and

2.3.3, the current–field characteristics under direct voltage were probably determined, not by mechanisms in the bulk of the liquid, but by space charges etc. in the vicinity of both the electrode–liquid interfaces.

To avoid the troublesome space charge and time effects associated with direct stress measurements, MacFadyen and Helliwell (1959), and Watson and Sharbaugh (1960), developed techniques for conduction studies using short-duration voltage pulses. An added advantage was the fact that the electric strength of liquids under pulse voltages is usually about 25 per cent greater than for direct stress, which allowed measurements to be taken at very high applied fields. The currents in n-hexane under pulsed stress were orders of magnitude greater than the currents of a few nA measured under static stress by Zaky *et al.* (see Fig. 2.11). For instance, MacFadyen and Helliwell measured about 10$\mu$A at a stress of 20 MV m$^{-1}$ whilst Watson and Sharbaugh detected nearly 3 mA at 130 MV m$^{-1}$. A possible physical explanation for the large difference between the steady and transient currents has already been outlined in sub-section 2.3.3. On the other hand, we have also noted in sub-section 2.3.3 that Kahan and Morant (1963) observed a current of several $\mu$A in highly-degassed n-hexane under a direct stress of 20 MV m$^{-1}$. This is comparable to the result of MacFadyen and Helliwell (1959). Thus, it may be that the currents under pulsed stress represent a sample of the large, 'noisy' fluctuations that are superimposed on the average current under static stress.

Watson and Sharbaugh (1960) extended their pulse measurements in a deliberate and systematic attempt to observe an $\alpha$-process in n-hexane. Fig. 2.13 shows the results obtained with flat stainless steel electrodes, at several gap lengths. As may be seen, the currents remained constant for fixed values of applied field, up to a stress of 124 MV m$^{-1}$, but there was a slight indication of an incipient $\alpha$-process at a gap of 125 $\mu$m and fields of 128 MV m$^{-1}$ and 132 MV m$^{-1}$. These fields were in the region of 90 per cent of the electric strength of n-hexane under pulse voltages (section 3.10). Consequently, Watson and Sharbaugh concluded that electron multiplication via collision ionization was not a significant source of current in liquids. However, this conclusion must be modified in view of the recent estimates by Devins and Wei (1972) of electron-trapping distances in n-hexane. A trapping distance of $\sim 1$ $\mu$m (see Table 1.4.3) is less than 2 per cent of the smallest gap length used by Watson and Sharbaugh (1960). Thus, many of the electrons were probably captured by impurity molecules within a short distance from the cathode so that the number capable of ionizing collisions was reduced considerably. Under these circumstances an $\alpha$ coefficient would be very small, as was found by Watson and Sharbaugh.

Gallagher (1968) has obtained a fairly reliable indication for an $\alpha$-process in liquid argon. A 1 $\mu$Ci plutonium source deposited on the cathode provided radiation-induced currents, which were measured at several gaps, as shown in Fig. 2.14. The small increases in current, denoted by the dotted lines in Fig. 2.14,

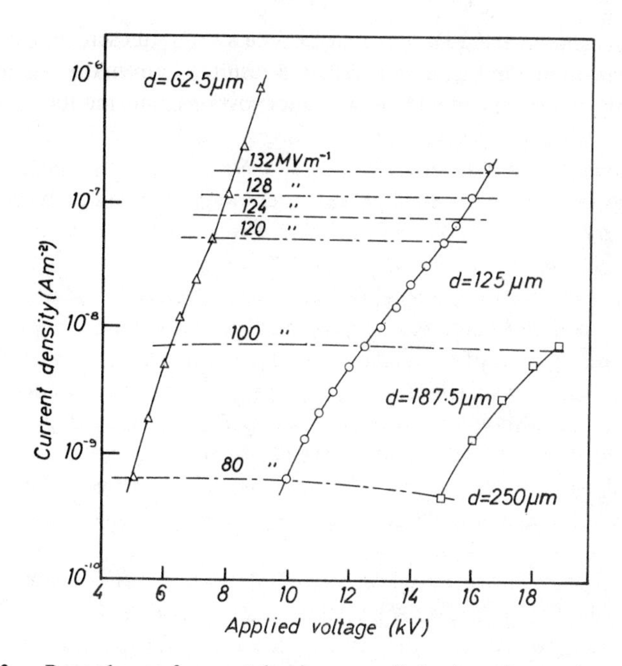


FIG. 2.13. Dependence of current density on applied pulse voltage. n-hexane between stainless steel electrodes. Pulse durations of 5, 3, and 1.5 $\mu$s (after Watson and Sharbaugh 1960).

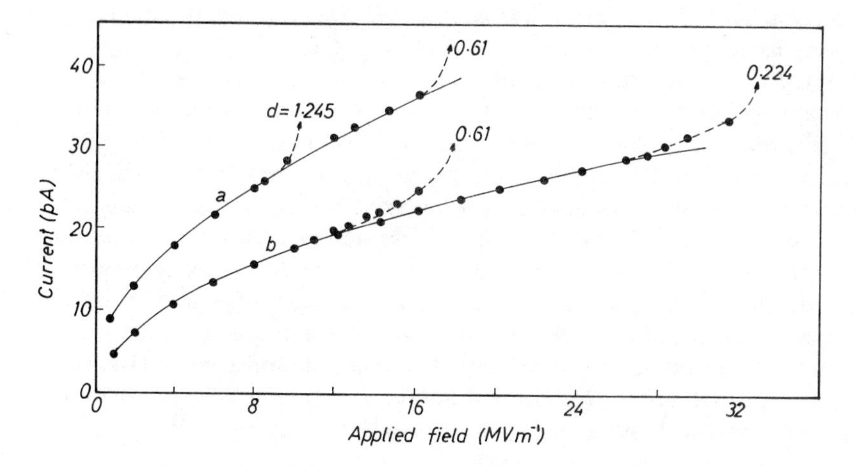


FIG. 2.14. Variation of radiation-induced current with applied field in liquid argon. Gap lengths (mm) indicated on each curve. The arrows indicate that breakdown occurred before a current reading was taken at that applied field (after Gallagher 1968).

were taken as evidence that an $\alpha$-process was present. The decrease in current from curve *a* to curve *b* in Fig. 2.14 resulted from discharges in the test cell which were unpredictable and which, over some months, reduced the activity of the radioactive cathode. However, by restricting the applied field to about 90 per cent of the expected breakdown value, the measurements of curve *b* in Fig. 2.14 were repeated several times. Identical results were obtained each time. On the basis of eqn (2.3), an estimate of $10^{-1}$ was obtained for $\alpha d$ at stresses just below breakdown. With the aid of a pulse-height analyser an investigation was made of the dependence on gap length of the size of the current pulses produced by individual radioactive emissions. The results confirmed that $10^{-1}$ was an upper limit to $\alpha d$. In view of the small value for $\alpha d$ in liquid argon, where electrons have high drift velocities, it is not surprising that there was little sign of an $\alpha$-process in n-hexane, where many electrons were trapped by impurities. It is interesting to note that the theory of breakdown in liquids proposed by Swan (1961) yielded values for $\alpha d$ between 0.1 and unity when applied to breakdown measurements in liquid argon. We shall consider this theory in some detail in section 3.5.

Many investigators have rejected completely the concept of a Townsend process in liquids. The arguments were based largely on the fact that liquid densities are about $10^3$ times higher than for gases and consequently, from the kinetic theory of gases, an electron mean-free-path should be much smaller in a liquid. A gas at atmospheric pressure has a mean-free-path in the region of $10^{-7}$ m. Neighbouring molecules in a liquid are separated by distances $< 10^{-9}$ m and it was assumed that electron mean-free-paths were of this order. With an applied field of $10^8$ V m$^{-1}$, which is close to the electric strength of a liquid, an energy of $10^{-1}$ eV can be gained from this field over a distance of $10^{-9}$ m. This energy is too small to cause ionization. However, it is not justified simply to extrapolate on the basis of kinetic theory. We have seen in section 1.4 that electron mobilities can differ by several orders of magnitude in liquids which possess comparable densities. Thus, electron mean-free-paths in liquids may be much longer than $10^{-9}$ m. Moreover, an $\alpha$-process is present in solids which exhibit high densities: ionization in the depletion layer of a *p–n* semiconductor junction is a well-established fact. Since an electron multiplication process occurs in gases and in solids, it is not unreasonable to assume that it can happen in liquids, although the collision mechanisms may be quite different in the different stages of condensation. Positive evidence for collision ionization may be obtained by repeating the experiments of Watson and Sharbaugh (1960) in a liquid which can sustain high electron mobilities. A suitable medium might be argon or some of the ultra-pure hydrocarbons listed in Table 1.4.1.

## 2.4. Summary

Ionic impurities are chiefly responsible for conduction in insulating organic liquids at low and at intermediate fields. At this stage it is difficult to decide if molecular complex formation can influence the current in aromatic hydrocarbons. Besides space charges, the slow decay of current with time under d.c. fields may be caused by a gradual egression of solid particles from the liquid between the electrodes. At high fields, however, particle contributions to the total current flow are small. Electrode effects dominate the current-controlling process and, in air-saturated liquids at high d.c. fields, the anode may exercise a much greater control than previously assumed. There is some evidence of ionization by electron collisions in liquid argon. Perhaps, in the future, collision ionization will be observed in hydrocarbons by means of conduction experiments on super-pure liquids. We await with interest the results of these experiments.

# 3

# Breakdown

## 3.1. Introduction

During the past fifty years many aspects of breakdown in dielectric liquids have been extensively investigated and there now exists a large amount of published data on the subject. However the measurements of many independent workers show large variations and often appear to be in complete disagreement. As a result there is a diversity of opinion on the breakdown process and several theories are currently in vogue, each offering different explanations for the mechanism of breakdown. In this chapter it is intended to examine carefully the various factors which can influence the breakdown voltage and, in this way, to suggest reasons for the apparent contradictory nature of some observations. Furthermore it will be indicated which theory seems most appropriate to explain the experimental results.

To avoid monotonous repetition the ratio of breakdown voltage to electrode separation will be referred to by terms such as electric strength, dielectric strength, breakdown strength, or simply, strength.

## 3.2. Experimental parameters affecting breakdown

The experimental parameters which can influence the result of a breakdown measurement in a liquid are numerous and may be classified under headings such as:

(*a*) test procedure;
(*b*) material or surface state of each electrode;
(*c*) field-configuration, area and separation of the electrodes;
(*d*) physical purity of the liquid;
(*e*) chemical purity of the liquid;
(*f*) temperature and hydrostatic pressure;
(*g*) duration of the applied voltage.

The parameters are not listed in order of importance as their individual effects may be changed by altering the experimental conditions. However, in order to obtain reliable data, it is essential to control these variables as much as possible. Furthermore, it is obvious from the foregoing remarks that the results of one person should only be compared with those of another when proper account is taken of the conditions pertaining to their experiments.

### 3.3. General considerations of modern breakdown theories

Before discussing the factors which can affect the breakdown process it is necessary to summarize briefly the hypothesis of each theory of breakdown in order to appreciate its relevance to the experimental results.

Early attempts to explain the breakdown of liquids were qualitative in nature and involved processes known to occur in gaseous breakdown. It was assumed that electron emission from the cathodes together with electron multiplication via collision ionization in the bulk liquid, operated simultaneously to produce a current instability leading to breakdown. These ideas of an $\alpha$-type Townsend process, as in gases, were modified and put on a quantitative basis by Lewis (1956), and later by Adamczewski (1957), for hydrocarbon liquids. However, these workers pictured the ionization process as occurring via mechanisms, originally proposed by von Hippel (1937) for breakdown in ionic crystals, whereby the energy barrier to ionization was determined by electrons, emitted from the cathode, losing their energies through the excitation of molecular bond vibrations. Consequently, the electric strength of liquids should be markedly dependent on cathode material and liquid structure, and as we shall see, some correlations do, in fact, exist. Swan (1961) has also postulated that electron emission and charge multiplication were essential to the process of breakdown. By solving space charge and cathode emission equations for the condition of very small ionization he has developed a criterion for breakdown which yields fair agreement with experimental results, particularly when applied to liquefied argon. Models of this type have a number of shortcomings, however. In particular, they cannot explain the observed pressure dependence of the breakdown strength of liquids which indicates that a change of phase may be involved at a critical stage in the breakdown process. Moreover, they also predict the existence of an electron multiplication, or $\alpha$-process, for which there is scant experimental support (see sub-section 2.3.4).

Watson and Sharbaugh (1960) have considered the possibility of liquid breakdown as a thermal process. The idea is derived from the observation of large current densities near breakdown under impulse voltage conditions (sub-section 2.3.4). The current is assumed to originate from asperities on the surface of the cathode giving localized input energies which are thought sufficient to create a vapour bubble in the liquid. Once the bubble has attained a critical size, breakdown is believed to follow rapidly. The authors claim that a thermal model predicts the pressure, temperature, and molecular structure dependences of the electric strength of liquids, and they obtain good agreement with experimental results, particularly for the aliphatic hydrocarbons.

A 'bubble' mechanism has also been proposed by Kao (1960), by Krasucki (1966), and by Thomas (1973 *a*). Kao has developed an expression for the electric strength in terms of some of the physical properties of a liquid and the initial radius of the bubble. The agreement between theory and experiment was not very good and the theory did not take account of the production of the initial bubble. In Krasucki's theory, the presence of impurity particles, produced by spark-erosion of the electrodes, was invoked as a prerequisite for the creation of the initiating bubble. A mathematical treatment was given whereby a criterion for bubble growth and subsequent breakdown was obtained such that at the point of highest electrical stress, assumed to be in the vicinity of particles, the electromechanical pressure acting on the liquid should be at least equal to the sum of the hydrostatic and surface tension pressures. It was asserted that the theory was in quantitative agreement with measured values of the electric strength of liquids and could explain satisfactorily many of the hitherto controversial aspects of breakdown. Thomas (1973 *a*) adopted Krasucki's idea that breakdown is initiated by cavitation processes at the cathode. However, Thomas included several extra terms in his mathematical treatment of bubble growth, but he neglected completely any effect of particles. Despite this omission, the predicted dependence of strength on pressure, temperature, and gap length was in good agreement with previous experimental results.

A model of breakdown based on the presence of suspended polarizable particles in the liquid has been proposed by Kok and Corbey (1956, 1957 *a, b,* 1958). According to these authors breakdown was caused by the formation of bridges of particles between the electrodes: the bridges being formed by dielectrophoretic forces. Whilst the model is probably valid for commercial insulating liquids (Sharbaugh and Watson 1962) it is unlikely to be important in highly purified liquids. It also predicts an increase in electric strength with increasing temperature, which is contrary to experimental results (sub-section 3.9.2).

Finally, mention must be made of the statistical theory of impulse breakdown originally proposed by Saxe and Lewis (1955) and extended by Ward and Lewis (1960). Whilst the theory is not meant to explain the actual process of breakdown, it has highlighted the fact that measurements under impulse voltage conditions need to be assessed on a statistical basis. However, further treatment of this topic is deferred to section 3.10.

### 3.4. Dependence of electric strength on test procedure

Lewis (1959) has given a general description of the experimental techniques and equipment usually used for breakdown measurements. Automatic test equipment for breakdown experiments and for time lag measurements has been described by Brignell (1963) and by Metzmacher and Brignell (1968 *a*), respectively. Here, we shall only consider the influence on results of two different test procedures. The procedure normally adopted for breakdown experiments has involved a 'multi-shot' technique whereby the strength is determined from a long sequence of discharges on the same sample of liquid and pair of electrodes. The variation of the strength in such a sequence can usually be divided into three distinct regions, as may be seen in Fig. 3.1. The strength is found to increase (or decrease) with the initial few breakdowns (region *a*) to a steady level which can exist for an extended number of breakdowns (region *b*), after which it deteriorates into a wild and erratic behaviour (region *c*). Usually, conditioning, as it is called, to a higher strength is observed but a downward trend has also been found. The conditioning process has been attributed to the removal of adsorbed gases from the electrode surfaces (Maksiejewski and Tropper 1954), but it has also been found in degassed liquids with degassed electrodes (Watson and Higham 1953). It has also been attributed to the removal from the cathode of micro-bubbles (Hancox and Tropper 1957) and of projecting, highly emissive sites, and also to the degree of filtration to which the liquid has been subjected to remove impurity particles (Salvage 1951). More recently, it has been observed that in liquid nitrogen even the anode material may influence the upward or downward trend in the conditioning period (Keenan 1972), but the reasons for this are still obscure. A more detailed discussion of conditioning is to be found in the article by Lewis (1959).

Whilst the discharges during conditioning may remove active areas from the cathode, each additional spark will produce a crater on the electrodes and gas evolution in the liquid, so that the extent of the stable period of measurements will be determined by a combination of damage to the electrodes, decomposition products, and impurity particles in the liquid. For direct voltage breakdown in organic liquids and liquefied gases, the use of rapidly-acting electronic circuits to

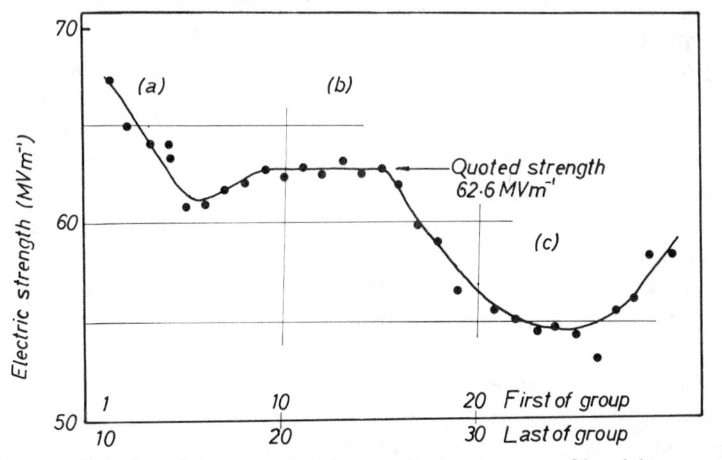

FIG. 3.1. Variation of the mean electric strength in a sequence of breakdown measurements in n-hexane, at a gap length of 200 μm and a temperature of −10°C. Each point is the mean of 10 consecutive measurements, that is, 1-10, 2-11, etc. in the sequence (after Brignell and House 1965).

suppress the spark discharge energy (Saxe and Lewis 1955) has allowed workers to make several hundred measurements on a single sample of liquid and pair of electrodes without much lowering of the average strength. However, the measurements on organic liquids were undoubtedly influenced by dissolved air or oxygen which had a stabilizing effect on the strength (section 3.7) since in oxygen-free n-hexane the steady period is limited to some twenty or thirty breakdowns (region *b* in Fig. 3.1). The rapid decline in strength after the steady period is of little scientific interest and can be associated with the excessive damage to electrodes and liquid caused by the preceding discharges.

Rather than use a multi-shot procedure where, as already stated, each discharge will inevitably produce a change in the experimental conditions, several workers have adopted the procedure of using fresh electrodes and fresh liquid for each measurement. However, great care is needed in any comparison of results obtained by either method, since the result from a single-shot experiment should only be compared with the result of the first breakdown in the conditioning period. In the past, conditioning has largely been ignored but a proper appreciation of the significance of the initial breakdown and the conditioning period could clarify much of the conflicting evidence which has appeared in the literature. This is especially true in the interpretation of impulse measurements where these two different procedures have mostly been employed (section 3.10).

The magnitude of the initial breakdown can be associated with the peculiar surface properties of freshly prepared electrodes. To obtain reproducible surface conditions the most common method used is a mechanical polishing technique whereby surface irregularities are removed by buffing the electrodes on a high-speed mop to which has been applied a polishing compound appropriate to the particular metal. With a little skill and abundant patience it is possible to produce an electrode with a mirror-like, nominally scratch-free, surface. However, little attention has been paid to the effect of the polishing process itself or to the type of surface layer that is produced. It is well-established (Grunberg and Wright 1953, 1955) that polished surfaces have properties different from those of the bulk metal and that they consist essentially of a thin layer of extremely fine metal crystals interspersed with particles of oxide produced at the high temperatures prevailing during polishing. Also, these temperatures are sufficient to provide the activation energies necessary for chemisorption and adsorption of gases on the surfaces. Even for gold there is strong evidence to suggest that the polishing process produced a surface layer with very different properties from those of the metal (Gallagher and Lewis 1964 *a*).

The phenomenon once denoted by the term 'exo-electron emission' is another factor which may have a bearing on the magnitude of the initial breakdown (cf. the article by Grunberg 1958 for a discussion of exoemission). It has been observed that freshly abraded specimens of aluminium, copper, brass, and nickel when introduced into a Geiger counter have triggered off a large number of pulses. This high activity was explained in terms of electron emission associated with ionization phenomena accompanying the reaction of oxygen with the deformed metal surface. From these remarks it may be concluded that freshly prepared electrodes can exhibit an anomalously large electron emission and a high degree of surface activity. Therefore using a multi-shot procedure an upward trend in the electric strength is to be expected and conditioning may proceed by any or all of the mechanisms already listed. The downward trend has only been found when exceptional care was taken to produce scratch-free surfaces. As this was likely to involve a protracted period of polishing it is probable that the surface was coated by a layer of oxide. On the basis of the above arguments this surface must be classified as initially very inactive, with a tendency to expose active sites and produce lower strengths as the number of breakdowns is increased. It is reassuring to find that an oxide layer can usually reduce the emission properties of an electrode (see curve C in Fig. 2.11 and section 3.5).

### 3.5. Influence of electrode material

Many workers have looked for a cathode dependence of the strength of organic liquids but more contradictory results have been obtained for this effect than any other. The possibility that the anode might alter the strength was rarely considered, despite the fact that the anode was known to play a role in the breakdown process in vacuum (Denholm 1958) and air (Germer 1959) at small electrode spacings. It is only recently that tests on liquefied gases have demonstrated that the anode metal can control the strength of a liquid; a control which is probably present in organic liquids but is masked by discharge products on the electrodes. We shall deal firstly with results which were interpreted in terms of cathode effects alone.

For many theories (section 3.2) the emission of electrons into a liquid is considered to be one of the primary processes of breakdown. This implies a dependence of the electric strength on the work function of the cathode metal, but the evidence from the literature indicates that it is extremely difficult to obtain any definite relationship. For example, in n-hexane with direct voltages, Salvage (1951) found some correlation between strength and work function for a range of metals, while Maksiejewski and Tropper (1954) observed a distinct dependence on the metal used which disappeared when the electrodes and the liquid were degassed. Maksiejewski and Tropper (1954) found that the strength increased with electrode metal in the order $\mathrm{Ag} < \mathrm{Pt} < \mathrm{Al} < \mathrm{Cu} < \mathrm{Fe} < \mathrm{Ni}$ whilst the work function of these metals when not degassed increased in the order $\mathrm{Al} < \mathrm{Ag} < \mathrm{Pt} < \mathrm{Ni} < \mathrm{Cu} < \mathrm{Fe}$. On the other hand, Lewis (1953), using similar techniques in n-hexane, found no significant differences between Al, Cu, Pt, and stainless steel electrodes. When organic liquids are tested with impulse voltages the situation is again confused. Using fresh electrode surfaces for each measurement Goodwin and MacFadyen (1953) found a definite dependence on work function for n-hexane and methyl alcohol, but, in later work on n-hexane by Holbeche (1956) and on transformer oil by Zein El-Dine and Tropper (1956), the electrode effect was negligibly small. As outlined in section 3.4, a metal that is mechanically polished is most unlikely to exhibit a work function corresponding to its photo-electric value and it is not surprising that conflicting results have been obtained. Moreover, in view of the measurements of conduction currents in sub-section 2.3.2, a clear dependence on work function would not be expected in break-down experiments.

As already mentioned, the conflicting evidence of many workers concerning the effect of the cathode material on the strength of organic liquids has been attributed to variations in the cathode surfaces used by them. However, other factors not previously considered could have been responsible for these differences. Especially, the results do not preclude the possibility of an anode influence on the strength since it was not investigated, nor was account taken of the probable differences in air content in each liquid (section 3.7). Some of the confusion could also arise because, under direct stress, layers due to pre-stressing (Ward and Lewis 1963) will form and mask the true metal properties, but under impulse conditions this is less likely, and differences between metals will be more apparent.

In order to investigate electrode effects more precisely, it was recognized that measurements should be made on a type of liquid where the deleterious products of discharges would be reduced to a minimum. An obvious choice was the liquefied gases. Unlike hydrocarbons and oils a spark in these liquids results only in the liberation of a gas bubble which ultimately condenses again. Of this group liquid argon could be considered as the ideal test liquid, as it does not react with either the electrodes or the impurities, and deposits on the electrodes are avoided completely. Argon gas is obtainable in an extremely pure state, at a moderate price, (99.995 per cent pure with approximately 100 p.p.m. of nitrogen and 20 p.p.m. of oxygen as the main impurities) and is easily condensed using liquid nitrogen as the refrigerant. The very low temperatures associated with the liquefaction of helium and neon create experimental difficulties for breakdown studies and they are both seriously influenced by quantum effects, while krypton and xenon are expensive to obtain in a pure state. Both liquid oxygen and nitrogen can be easily produced but the oxygen is highly reactive with many metals, whilst in liquid nitrogen, containing oxygen, oxides of nitrogen could be formed. A compilation is given in Table 3.5.1 of the values of d.c. electric strength for all cryogenic liquids in which breakdown measurements have been made. A pronounced electrode influence on the strength is obvious, some values changing by almost 70 per cent with a change in electrode material. However, there is no correlation with work function of the cathode metal. In fact, for the inert liquid argon the strength is reduced as the work function is apparently increased through the use of copper, gold, and platinum cathodes respectively. Despite this lack of dependence on work function, quite remarkable

Table 3.5.1.

*Electric strengths of cryogenic liquids as a function of electrode material*

| Electrodes | Reference | Electric strength† (MV m$^{-1}$) | | | | | |
|---|---|---|---|---|---|---|---|
| | | Helium | Hydrogen | Nitrogen | Argon | Oxygen | Methane |
| Stainless steel | (1) | – | – | 100 | 100 | 104 | – |
| | (2) | – | – | 180 | 140 | 238 | – |
| | (3) | 105 | – | – | – | – | – |
| | (4) | 40 | 134 (60Hz) | 80.5(60Hz) | – | – | – |
| | (5) | 70 | – | – | – | – | – |
| | (6) | 70 | – | – | – | – | – |
| | (7) | – | – | – | 190(1.5μs pulse) | – | – |
| | (8) | – | – | – | – | – | 153 |
| | (9) | – | – | 146 | – | – | – |
| Brass | (2) | – | – | 162 | 101 | 144 | – |
| | (8) | – | – | – | – | – | 136 |
| | (9) | – | – | 83 | – | – | – |
| | (10) | – | – | 70 (50Hz) | – | – | – |
| Copper | (2) | – | – | – | 140 | 181 | – |
| | (8) | – | – | – | – | – | 147 |
| | (10) | – | – | 104 | – | – | – |
| Gold | (2) | – | – | 150 | 116 | 124 | – |
| | (7) | – | – | – | 102 | – | – |
| Platinum | (1) | – | – | 93 | 86 | 93 | – |
| | (2) | – | – | 224 | 110 | 200 | – |
| Aluminium | (7) | – | – | – | 160(10μs pulse) | – | – |
| | (8) | – | – | – | – | – | 100 |
| | (9) | – | – | 114 | – | – | – |
| Tungsten | (11) | 72 | – | – | – | – | – |
| | (12) | 160(2.14K)<br>130(4.2K) | – | – | – | – | – |

† Quoted values of the strength refer to specific conditions. The values will change with a change in gap length, temperature etc. References are identified by: (1) Kronig and Van de Vooren (1942); (2) Swan and Lewis (1960); (3) Blank and Edwards (1960); (4) Mathes (1967); (5) Fallou *et al.* (1969); (6) Gerhold (1972); (7) Gallagher and Lewis (1964*a*); (8) Sakr and Gallagher (1964); (9) Keenan (1972); (10) Kawashima (1974); (11) Blaisse *et al.* (1958); (12) Goldschvartz *et al.* (1972).

and reproducible electrode effects have been observed, particularly in argon, and we shall select this liquid for special consideration.

Fig. 3.2 shows the strength of argon under direct voltage as a function of the degree of oxidation of the electrodes measured as the time of oxidation from the cessation of the initial polishing (Swan and Lewis 1961). Since the noble metals each gave a constant strength, the rise to a peak for the oxidizing metals was interpreted in terms of a decrease in emission caused by the growth of a thin oxide layer for short oxidation periods. Thereafter, the fall in strength was attributed to an increased emission caused by positive ions acting across the insulating oxide layer on the cathode. This seemed a plausible explanation until it was discovered that the strength could be influenced by the anode metal also. By maintaining the surface conditions of either electrode constant it was possible (Swan and Lewis 1961) to separate the individual roles of the cathode

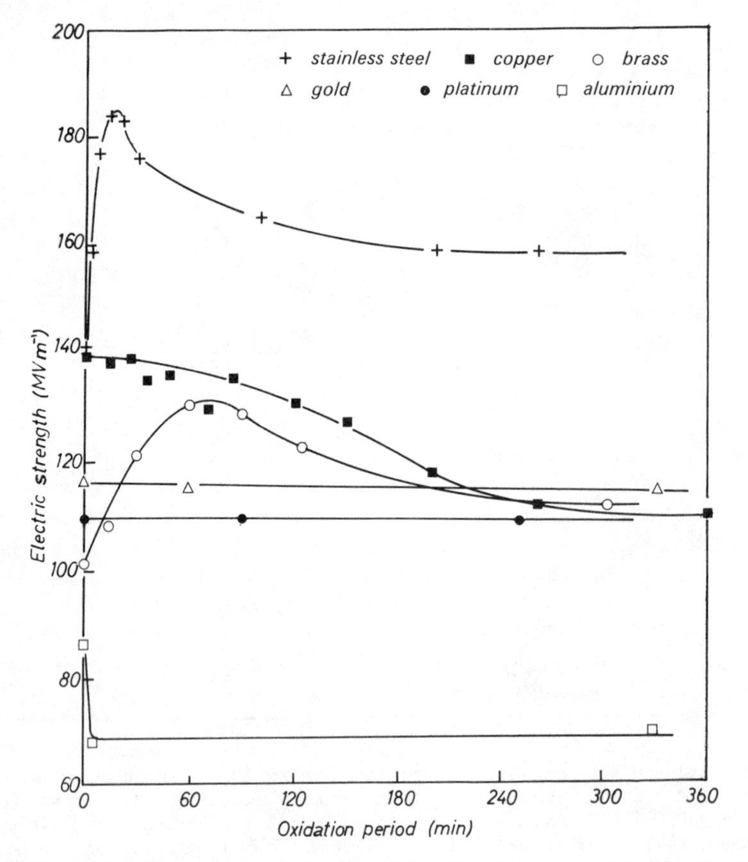


FIG. 3.2. Electric strength of liquid argon with oxidized electrodes (after Swan and Lewis 1961).

and anode and to show that the peaked behaviour in Fig. 3.2 resulted from the combined influences of both electrodes working in opposition (Fig. 3.3). This was the first time that the influence of the anode on the electric strength of liquids had been investigated in detail, although Sharbaugh, Cox, Crowe, and Auer (1955) had made a brief reference to the effect that the impulse strength of n-hexane was reduced by 9 per cent when the anode was roughened.

We have already considered in detail in sub-section 2.3.2 the mechanism proposed by Swan and Lewis to account for the unusual control of the anode on the breakdown of liquid argon. The suggested mechanism of anode control could be verified in two separate ways. Firstly, by using direct voltages and removing the trace amount of oxygen impurity, negative ions should not be formed and

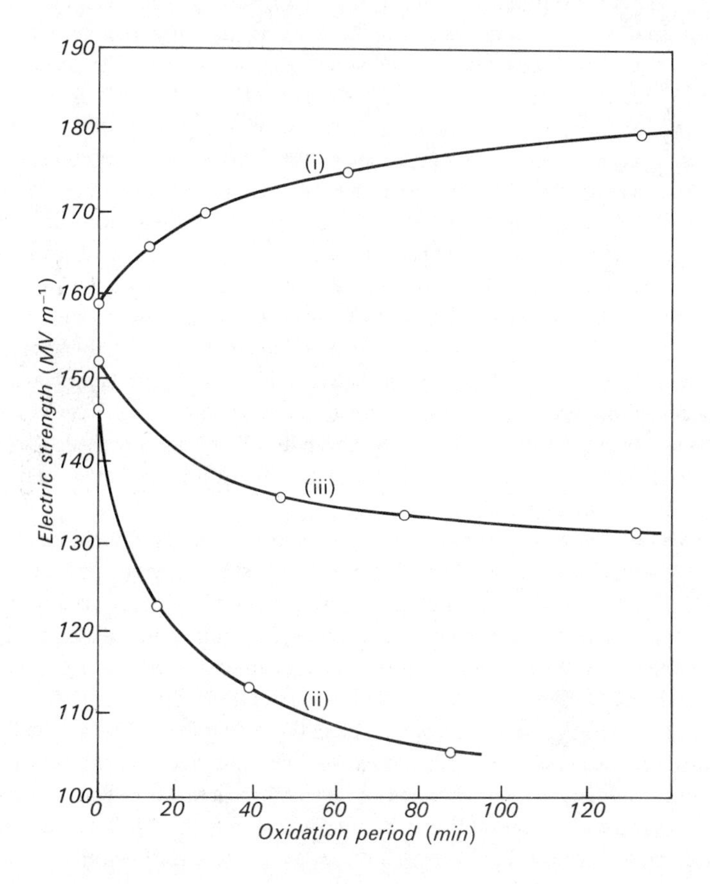


FIG. 3.3. Influence of anode and cathode on the electric strength of liquid argon. (*i*) cathode stainless steel of increasing oxidation time, anode stainless steel of 'zero' oxidation time; (*ii*) cathode gold, anode stainless steel of increasing oxidation time; (*iii*) cathode stainless steel of 24 hour oxidation time, anode stainless steel of increasing oxidation time (after Swan and Lewis 1961).

collected as an anode space charge and secondly, by using pulse voltages of sufficiently short duration, it should be possible to prevent the development of a negative space charge at the anode. As was expected, no anode effect was observed in the former case and neither an oxygen nor an anode effect in the latter case (Gallagher and Lewis 1964 *a*).

It is now possible to advance reasons why a dependence on anode conditions had not been previously observed in hydrocarbon liquids. In the first instance, there were few reported experiments of breakdown in these liquids where cathode and anode phenomena were studied separately. Both Edwards (1951) and Hancox (1957) found that the impulse strength of benzene and transformer oil was dependent on the cathode material only, but this is not surprising in view of similar results for liquid argon (Gallagher and Lewis 1964 *a*). Secondly, it is most probable that the anode was strongly blocking in all previous tests on hydrocarbons under direct voltage conditions. This would arise because of inherent impurities, probably of a wax-like nature (MacFadyen 1955), deposited on the anode and also because these liquids most probably contained minute quantities of oxygen (only a few parts per million of oxygen in liquid argon is necessary to produce the effect). Therefore, even gross changes in the underlying anode metal may be rendered ineffective and an anode influence on the strength, although present, would not be apparent from the results. It is very significant that similar anode effects (Fig. 3.4) have been observed in breakdown studies in liquid methane (Sakr and Gallagher 1964), which is the simplest of the hydrocarbons, and more recently in liquid nitrogen (Keenan 1972). Furthermore, some evidence for the establishment in n-hexane of negative ion layers at the anode has been obtained from prestressing experiments (Ward and Lewis 1963), while the anode mechanism proposed by Swan and Lewis has been adopted to explain the changes in the current–field characteristics obtained by several investigators (sub-section 2.3.2).

Since we have dealt at some length with breakdown measurements on liquefied gases it is appropriate, at this stage, to consider a theory of breakdown of liquids which has achieved some success when applied to results in liquid argon, and in which it is possible to incorporate, in a qualitative sense, both the effects of oxygen and the anode on the electric strength (Swan 1961). The theory was similar to earlier theories of Goodwin and MacFadyen (1953) and of O'Dwyer (1954) in that space charge at the cathode and ionization in the liquid were considered necessary for breakdown, but with the essential difference that the amount of ionization was assumed to be small, in agreement with experimental observations of conduction currents (sub-section 2.3.4). It was assumed that the electron current from the cathode occurred via a field-emission process and could be described by an equation of the form:

$$j_e = Am^2 E_c^2 \exp\left[-\frac{B}{mE_c}\right], \tag{3.1}$$

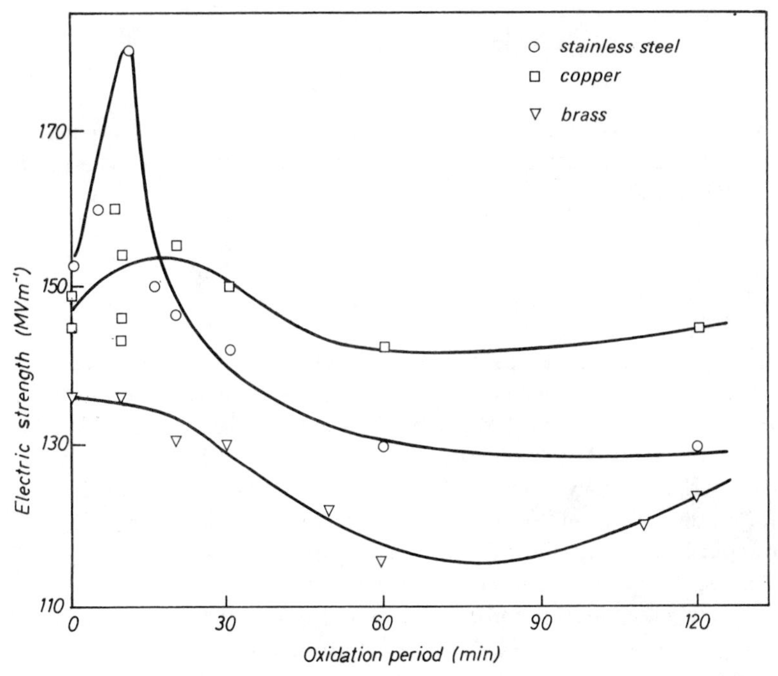


FIG. 3.4. Influence of electrode oxidation on the strength of liquid methane (after Sakr and Gallagher 1964).

where $E_c$ is the cathode field, $j_e$ is the electron current density at the cathode, and $A, m/$ and $B$ are constants depending on the cathode surface. The space-charge equation was solved for conditions of small ionization and it was shown that the cathode-field increment due to positive ions can be expressed as $E_+ = Gj_e$ where

$$G = \frac{2\pi d}{\epsilon\mu_+ E_0}\left[\alpha_0 d - \frac{\mu_+}{\mu_-}\right], \tag{3.2}$$

in which $d$ is the electrode spacing, $\epsilon$ the permittivity of the liquid, $\alpha_0$ is the value of the ionization coefficient for an applied field $E_0$, and the electronic and ionic mobilities are denoted by $\mu_-$ and $\mu_+$, respectively. Therefore for a positive field enhancement at the cathode the value of $\alpha_0 d$ must be greater than the ratio $\mu_+/\mu_-$, but $G$ can vary for different liquids since the ion mobilities may differ from one liquid to another. As it was well known that in liquid argon the ratio $\mu_+/\mu_-$ was about $10^{-4}$, which is negligible even for $\alpha_0 d$ as small as $10^{-2}$ in eqn (3.2). Swan considered this liquid ideal for a comparison between his theory and experimental results. In order to apply eqn (3.2) it was necessary to assume a functional form for the dependence of the ionization coefficient $\alpha_0$ with applied field $E_0$ as found by Ward (1958) for gaseous argon, namely

$$\alpha_0 = C \exp\left[\frac{-D}{E_0^{\frac{1}{2}}}\right] \quad , \tag{3.3}$$

where the constants $C$ and $D$ involve the gas pressure. Combining this equation with eqn (3.2) and neglecting the term $\mu_+/\mu_-$, we find that the cathode field $E_c$ satisfies the relation

$$E_c = E_0 + \frac{2\pi j_e C d^2}{\epsilon\mu_+ E_0} \exp\left[\frac{-D}{E_0^{\frac{1}{2}}}\right] . \tag{3.4}$$

The theory of Swan is best illustrated by considering Fig. 3.5. The curve $OF$ represents the cathode emission (eqn 3.1) and is due to ionization in the liquid and the subsequent movement of positive ions to the cathode. The applied field is given by the point $R_1$ and the cathode field, after a stable conduction current is established, is given by the point $T_1$. The situation at other applied fields $R_2$ and $R_3$ is also shown and it is obvious that a stable emission current is impossible if the applied stress is greater than the breakdown stress which is represented by $R_3$. This was taken to represent the breakdown condition. It was not possible to derive an explicit expression for the value of the applied field $E_0$ at breakdown but an explicit relationship for $d^*$, the gap at breakdown, was obtained which for our purpose may be written in the form

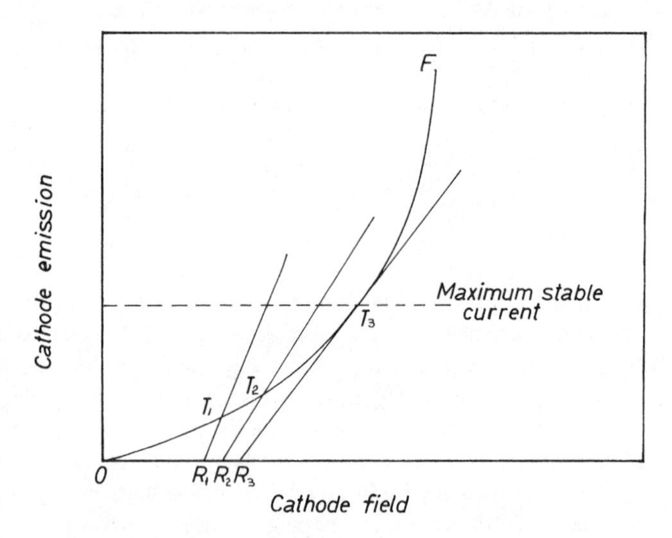


FIG. 3.5. Diagram illustrating the breakdown condition given by Swan (1961). The curve represents cathode emission (eqn. 3.1). The straight lines represent the cathode field–emission current relationship due to space charges (eqn. 3.4).

$$d^* = f\left[\frac{D}{E_0}, \frac{B}{E_0}\right]. \tag{3.5}$$

The reader is referred to the original article for an explanation of the function $f$. It is sufficient to note here that the first term in eqn (3.5) is linked with the ionization constant $\alpha_0$ and therefore represents liquid properties, whilst the second term contains the constants of the cathode. Obviously, depending on the relative magnitudes of these two terms the electric strength can be controlled either by the cathode or by the liquid properties. Eqn (3.5) was used to calculate the breakdown voltage–electrode spacing characteristics (Fig. 3.6) by choosing arbitrary but realistic values for the cathode constants and by assuming that the functional dependence of $\alpha_0$ on $E_0$ was as given in eqn (3.3). Although the

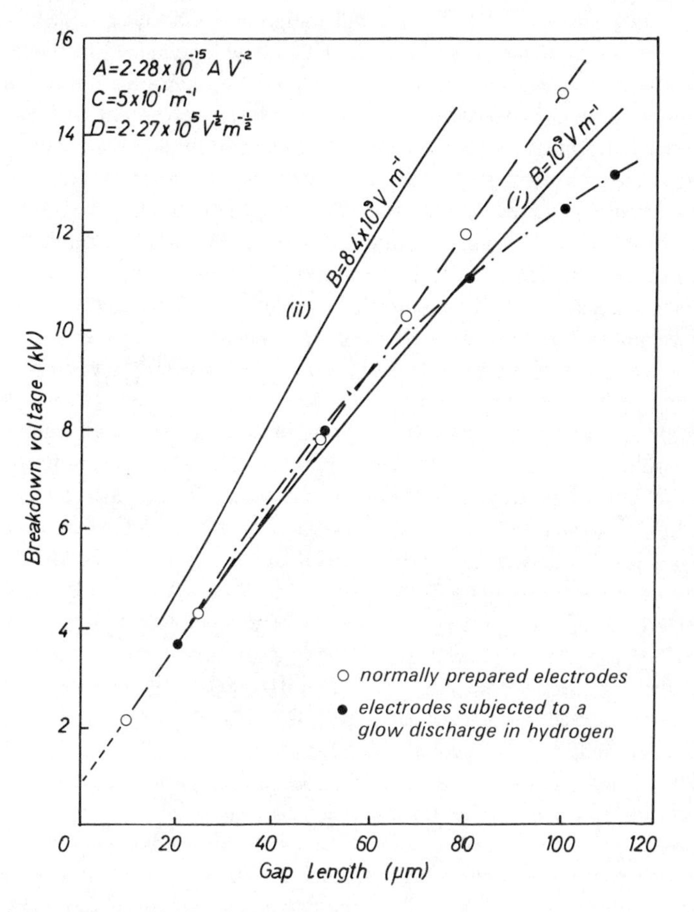


FIG. 3.6. Theoretical breakdown voltage–gap length characteristics compared with measurements in liquid argon (after Swan 1961).

agreement with experiment is good, it has been stated by Swan (1961) that the theory has limited application, in as much as it would be extremely difficult to determine some of the assumed parameters with any degree of certainty. However, some evidence for a small $\alpha$-type process in liquid argon has been deduced from measurements of induced ionization currents (see Fig. 2.14) and from experiments on its suitability as a spark-chamber medium (Riegler 1969). Furthermore, the theory can readily account for the anode influence on the d.c. strength of some of these liquids. If negative ions can form a space charge in a region close to the anode surface giving an enhanced field there, then a few extra ionizing collisions may occur. Moreover, if the anode field is sufficiently large, electron tunnelling from the liquid atoms into the anode may take place, as was suggested by Halpern and Gomer (1969 *b*) for their experiments on field ionization in cryogenic liquids. Each mechanism effectively increases $\alpha_0$. The current instability leading to breakdown will then occur at a lower value of $E_0$. Any increase in the blocking properties of the anode will enhance this space charge and lower the strength, and this is confirmed by experiments in argon with oxidized anodes (Fig. 3.3). In contrast to this qualitative agreement with experiment, we may note that the theory cannot be applied to liquid helium since $\mu_+/\mu_-$ is $>1$ (sub-section 1.3.4). Neither can it explain a pressure dependence on strength for helium, nitrogen, and hydrogen (sub-section 3.9.3).

Most other theories of breakdown have been concerned with explaining results for organic liquids and a full discussion of these will be given in later sections. Nevertheless, we shall now briefly consider two theories which appear the most relevant to liquefied gases. The boiling concept of Watson and Sharbaugh (1960) whereby breakdown follows the generation of a vapour bubble, might seem very appropriate to these low-temperature liquids, but it is unlikely to be suitable for several reasons. For instance, in the case of liquid helium, it would be expected that the large increase in its thermal conductivity on passing down through the $\lambda$-point temperature would help to increase its strength. However, the results in Fig. 3.7, obtained by Blank and Edwards (1960), indicate that HeII has a lower strength than HeI. A reduction in strength has also been noticed recently by Goldschvartz, van Steeg, Arts, and Blaisse (1972). Jefferies and Mathes (1970) have shown that much less energy is required to vaporize a given volume of hydrogen than is required for the same volume of nitrogen. This would suggest a large difference in their strength, which was not observed. Some objections can also be applied to the mechanism proposed by Krasucki (1966) involving the initiation and growth of bubbles in the vicinity of impurity particles in the liquid. Even though particles are probably present in cryogenic liquids it is likely that their density is much less than in organic liquids. If this were not the case 'natural' conduction currents (due to particle motion between the electrodes, as discussed in sub-section 2.3.1) in both types of liquid should be of the same magnitude, whereas liquefied gases possess negligible conductivities. Furthermore, in the theories of Watson and

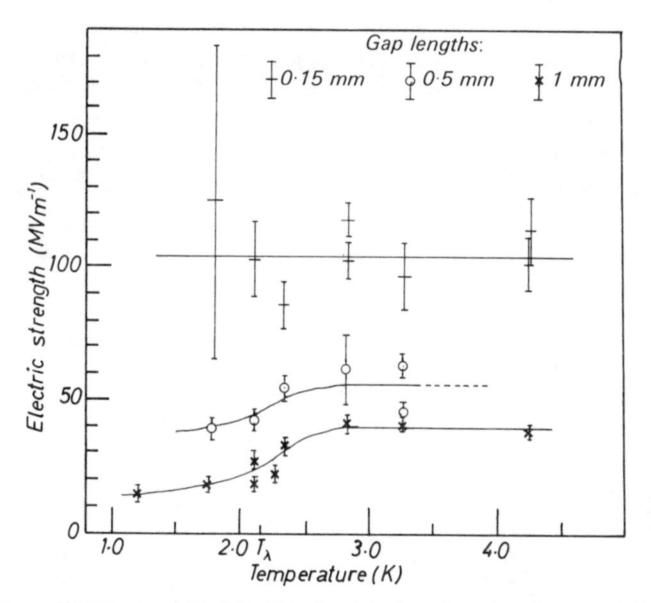


FIG. 3.7. Electric strength of liquid helium as a function of temperature (after Blank and Edwards 1960).

Sharbaugh (1960) and of Krasucki (1966), there is no provision for an oxygen nor an anode influence on the strength which, as we have noted already, has been found in argon, methane, and nitrogen. In fact, breakdown experiments on oxygen and helium liquids with mixed electrodes would probably reveal an anode control on their strengths. Electrons emitted into these liquids will, ultimately, migrate as stable negative ions, as we have observed in sub-section 1.3.4. A negative anode space charge, therefore, should easily be formed in oxygen and in helium. On the other hand, in liquid helium the negative ions provide a ready-made source of embryonic bubbles so that a discharge may develop along the lines suggested in Krasucki's theory. Much more work is needed to clarify the physics of breakdown in cryogenic liquids, particularly in helium.

## 3.6. Effect of electrode gap and geometry

So far we have discussed the influence of the metal and surface condition of the electrodes on the strength of liquids, but other factors such as the electrode gap (spacing) and geometry must also be considered. The dependence on gap has been the subject of much investigation, particularly because of the implications regarding the importance, or otherwise, of electron impact ionization in the events leading to breakdown. If ionization occurs, the strength should be dependent on gap and for pure liquids the strength generally increases as the gap is reduced. Fig. 3.8 shows the results of measurements on a paraffin series of hydrocarbons by Lewis (1953) which are typical of the results of other investigations in which either d.c. or impulse voltages were used. Lewis (1959) has adequately reviewed those earlier theories, based on collision ionization, which aimed at explaining results similar to those in Fig. 3.8. It is only necessary to comment here that the theoretical models were either too simple or yielded values of $\alpha$ which were too large.

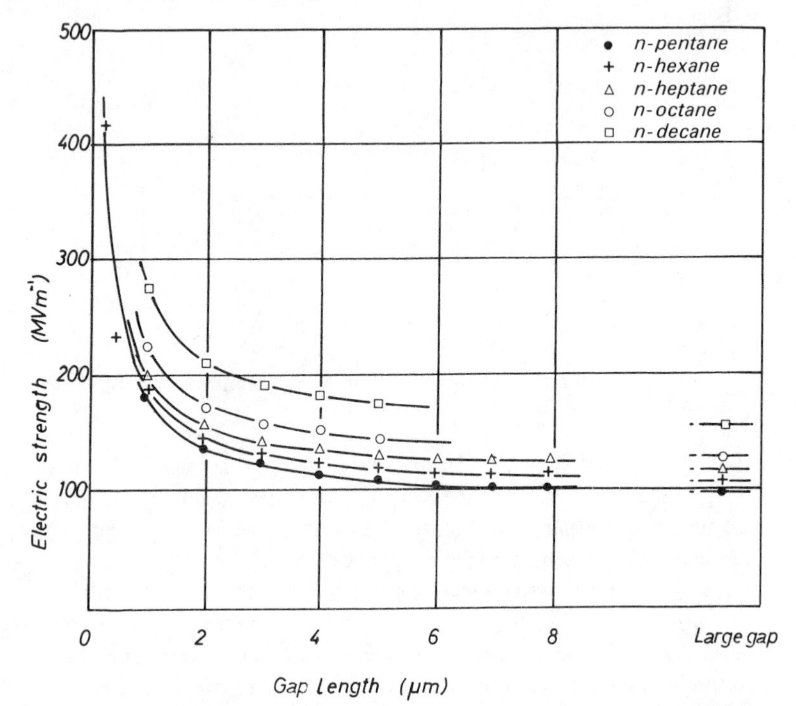


FIG. 3.8. Dependence of electric strength on gap length for paraffin hydrocarbons (after Lewis 1953).

An entirely different interpretation of electric strength–gap characteristics was proposed by Sharbaugh *et al.* (1955). It was argued that for spherical electrodes the area under electric stress increases with the product $Rd$ where $R$ is the electrode radius and $d$ is the spacing, and as most workers had used such electrodes, a decrease in strength with increasing spacing was inevitable, on the basis of an area effect. The effect of electrode area on the impulse strength of n-hexane was investigated using electrodes of five different radii, as shown in Fig. 3.9, and by normalizing these results with respect to the product $Rd$, a common curve was obtained, as in Fig. 3.10. The correlation between strength and electrode area was explained in terms of asperities on the cathode surface which contribute to the initiation of breakdown. This idea is probably correct,

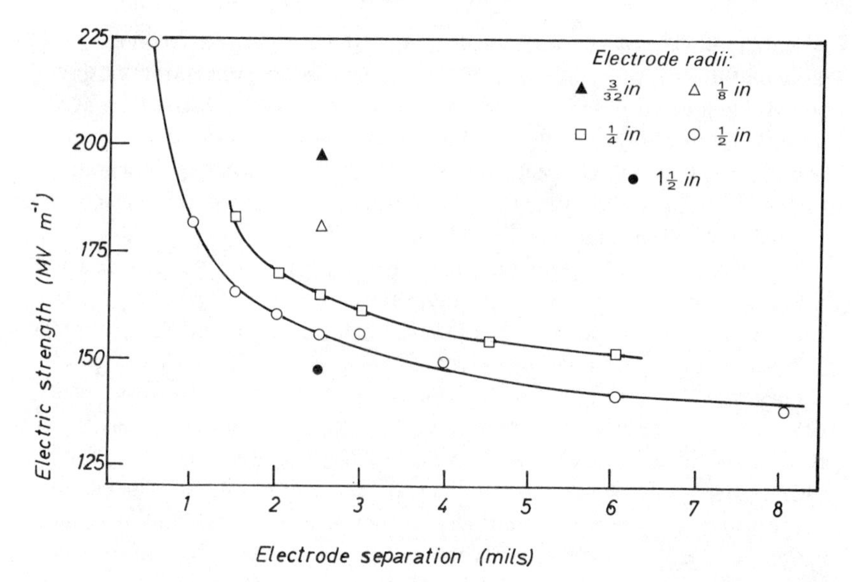


FIG. 3.9. Electric strength of n-hexane as a function of electrode separation (after Sharbaugh *et al.* 1955).

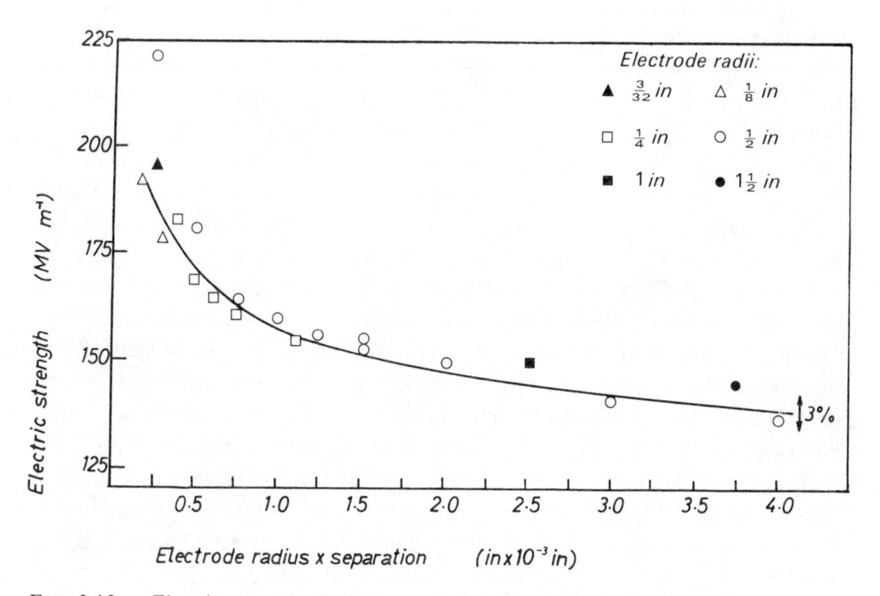


FIG. 3.10. Electric strength of n-hexane as a function of electrode radius times separation (after Sharbaugh *et al.* 1955).

particularly in view of the theoretical work of Metzmacher and Brignell (1968 *b*). Purely on the basis of a statistical model, and on the assumption that initiating events leading to breakdown occur at one of the electrodes, these authors have correctly calculated the variation of impulse strength as a function of spacing. More recently, Brignell and Metzmacher (1971) have shown, experimentally, that the mean rate of initiating events in n-hexane is proportional to electrode area, and that it is also independent of spacing.

Prior to about 1963 a common, yet fascinating, aspect of the breakdown voltage–gap characteristics was the intercept obtained on the voltage axis when these curves were extrapolated to zero gap. In fact the rapid increase in strength for gaps less than approximately 40 $\mu$m (Fig. 3.8) is a consequence of the intercept, which was in the region of 1000V for many liquids. However, recent work has shown that the characteristic of d.c. breakdown voltage versus gap can pass through the origin for gas-free n-hexane (Sletten and Lewis 1963), but yields an increasing intercept as the oxygen content of the liquid is increased. Similar examples have been found, such as transformer oil containing oxygen or nitrogen in solution (Gosling and Tropper 1964), whilst tests in liquid argon under d.c., and impulse, voltages gave a constant intercept of about 1000V (Gallagher 1962). Whether it is valid to extrapolate in this manner has recently been questioned by Hesketh (1966) who studied breakdown in the, hitherto, much neglected region of spacings less than 30 $\mu$m. The test cell, described elsewhere by Hesketh and Lewis (1969), had the advantage that measurements could be made on a single drop of liquid suspended between two wires, thus ensuring exceptional physical purity of the liquid. Fig. 3.11. shows the d.c. breakdown voltage plotted against gap for n-decane. The graph is linear down to a gap of 8 $\mu$m with an apparent intercept at zero gap of 940V, which value was observed to vary with a change in the material of the electrodes. However, point A in Fig. 3.11 is very significant in that at a spacing of 6 $\mu$m the breakdown voltage was increased by 14.1 per cent above the mean value. At smaller spacings a type of 'electromechanical' breakdown was evident, whereby the electrostatic forces of attraction between the wires were sufficient to reduce the spacing at a rate faster than the increase in opposing tensile forces. As a result the breakdown voltage decreased with spacing. Similar behaviour was found for n-hexane. Under pulses of microseconds duration a curve of the same shape as Fig. 3.8 was obtained and the breakdown characteristic was linear down to a spacing of 2 $\mu$m, presumably because mechanical movement of the wires did not occur in the short time available. The results of Fig. 3.11 support the proposal by Lewis (1953) and by Swan (1961) that the 'intercept phenomenon' could arise from a Paschen minimum, similar to that in gases: because of collision ionization in a gas a minimum is observed when the breakdown voltage is plotted against gap length. It is most intriguing to note that collision ionization can probably account for the observed increase in strength with decreasing thickness of alkali halide crystals (O'Dwyer 1973). However, Hesketh (1966) was more inclined to

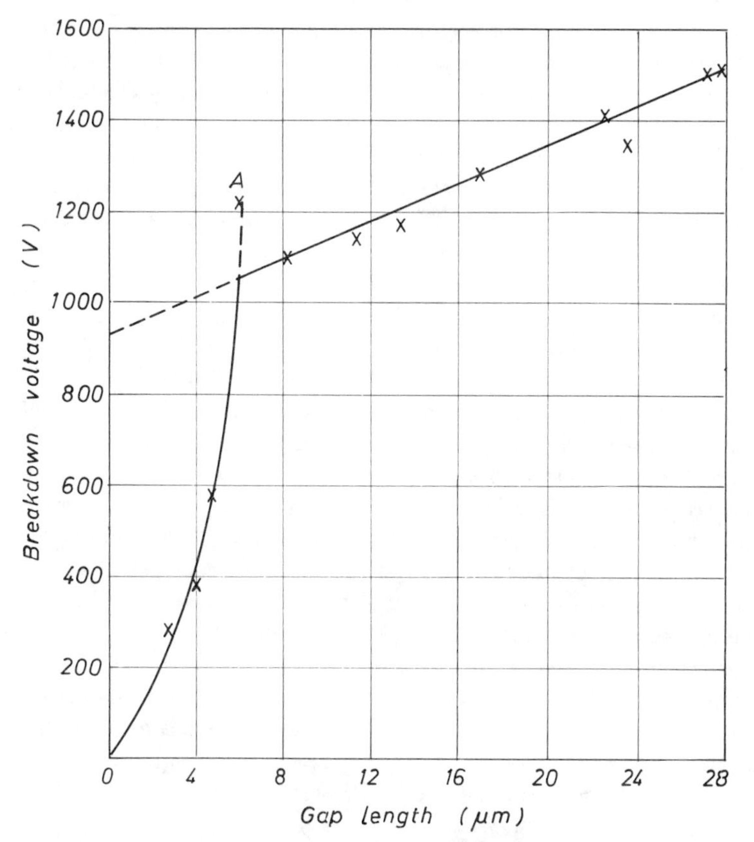


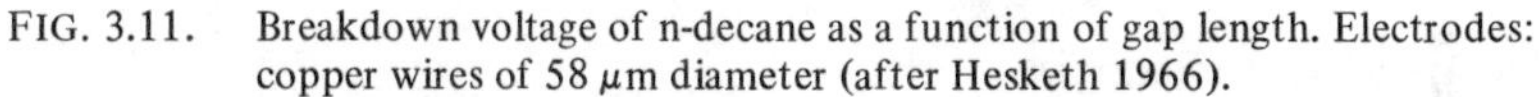

FIG. 3.11. Breakdown voltage of n-decane as a function of gap length. Electrodes: copper wires of 58 μm diameter (after Hesketh 1966).

the view that the high value of breakdown voltage at *A* in Fig. 3.11 reflected the fact that the liquid was apparently free from solid particles (section 3.8). Moreover, since the intercept appears to be independent of time (Gallagher 1962) but dependent on dissolved gases (Sletten and Lewis 1963), it is likely to be associated with oxide or gaseous layers at the electrode–liquid interfaces and may be a measure of the voltage drop across these layers (Gosling and Tropper 1964). Much more work is needed before this question can be resolved and, provided that the experimental difficulties can be overcome, measurements at very small spacings could yield valuable information about the mechanisms of breakdown. It may also be mentioned in passing, that some workers quote values for the electric strength which are slopes of the graphs of breakdown voltage–gap characteristics, whilst others express it as the ratio of average breakdown voltage to gap, taken at a fixed gap, so that some care is needed in any comparison between values which are given in the literature.

Nearly all breakdown measurements have been made with sphere–sphere electrode systems but point-sphere geometries can be a very useful tool for investigational purposes. With a negative point it is now generally accepted that the strength is independent of the cathode material and that breakdown is initiated via a copious discharge of electrons from the point (Chadband and Wright 1965). Once emitted, these electrons can interact with the liquid molecules and the breakdown voltage is determined by the ease with which the initiating discharge can propagate across the gap, or in other words, by the energy-absorption properties of the liquid. Consequently, for simple liquids, it is to be expected that the voltage–gap characteristic for negative point polarity will be lower than for positive point since, in this case, a much higher field is required to produce an initiating discharge from the spherical cathode, capable of leading to breakdown. Experimental results for pure hydrocarbons and liquefied gases confirm these predictions. However, the situation is complicated for more complex liquids, such as transformer oil which contains a mixture of compounds. The breakdown characteristic for the two polarities of the point can cross over at small or large gaps depending on the concentration and energy-absorption properties of the individual constituents of the liquid (cf. Zaky and Hawley 1973 for a discussion of these effects in transformer oil). It must also be remembered that, under d.c. stresses, copious electron emission from a negative point may provide currents in a liquid sufficiently high to cause significant enhancement of the field in the region of the sphere anode, as already discussed in section 3.5. Therefore, even with this polarity, some anode control on the breakdown voltage may be present, as was observed recently in liquid $N_2$ (Keenan 1972) and liquid Ar (Gallagher and Lewis 1964 *a*).

### 3.7. Effect of gaseous impurities

In Chapter 2 we have seen that gaseous impurities, particularly oxygen, can affect conduction currents in liquid hydrocarbons. It is hardly surprising to find that the electric strength is also influenced by trace amounts of dissolved gases. However, the importance of dissolved oxygen in measurements of breakdown voltage was not realized until relatively recently (Sletten 1959) although it was well known that it could alter the strength of other pure gases, and that it had a marked influence on the transport of electrons in liquid argon (Davidson and Larsh 1950).

Sletten and Lewis (1963) have reported a systematic investigation of the effect of oxygen, nitrogen, carbon dioxide, and hydrogen on the electric strength of n-hexane. Their results have shown that without oxygen in solution its d.c. strength is only 80 MV $m^{-1}$ but that when saturated with air the strength of n-hexane increases to about 120 MV $m^{-1}$. Tests showed that dissolved nitrogen had no such effect so that the increase in strength with dissolved air was due to the oxygen present. Hydrogen and carbon dioxide had no effect even when large amounts were dissolved, excepting that repeated discharges when $CO_2$ was present caused the strength to rise, probably because of the release of oxygen. Fig. 3.12 shows the increase in the d.c. strength of n-hexane with increasing oxygen and air concentrations, together with the dispersion of results. Above an oxygen content corresponding to an equilibrium partial pressure of about 130mm Hg, the strength 'saturated' at a value of about $125 \mathrm{MVm}^{-1}$. This value is within the range $110 \mathrm{MVm}^{-1}$ to $140 \mathrm{MVm}^{-1}$ which many previous investigators had obtained for the strength of n-hexane, even after attempting to degas the liquid. However, it is most unlikely that their degassing techniques were rigorous enough, especially when we see the extensive purification procedures which are essential for the removal of oxygen from n-hexane (Minday *et al.* 1971). By using degassing and distillation techniques, which were more elaborate than those used by Sletten and Lewis (1963), a strength of only $54 \mathrm{MVm}^{-1}$ has been measured by Brignell and House for n-hexane between nickel electrodes at a gap of 200 $\mu$m in 1965. From the foregoing remarks it must be concluded that results obtained for organic liquids prior to 1963 refer to liquids which contained significant amounts of dissolved oxygen. It is obvious now that many apparent discrepancies between previous measurements can be attributed to differences in the technique of liquid preparation and therefore to variations in the concentration of dissolved gas, notably oxygen. Also, these observations lend support to our assertion in section 3.5 that an anode effect arising from dissolved oxygen, although not apparent, was present in earlier measurements on hydrocarbons.

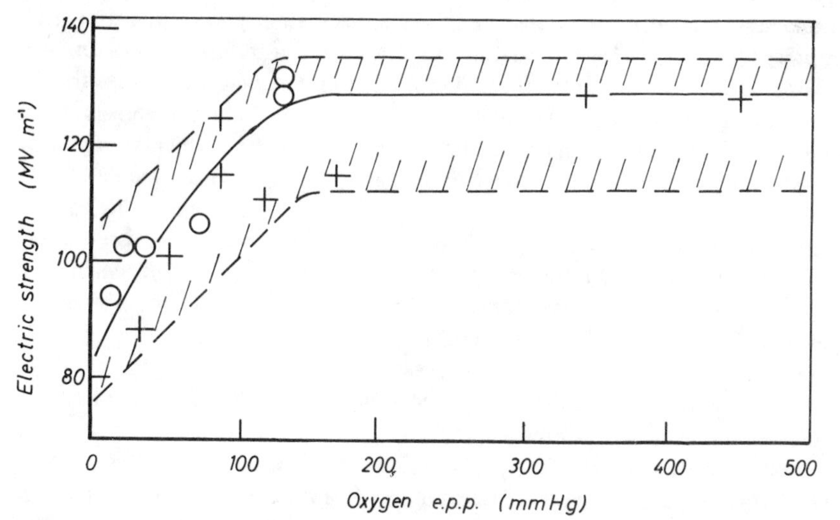


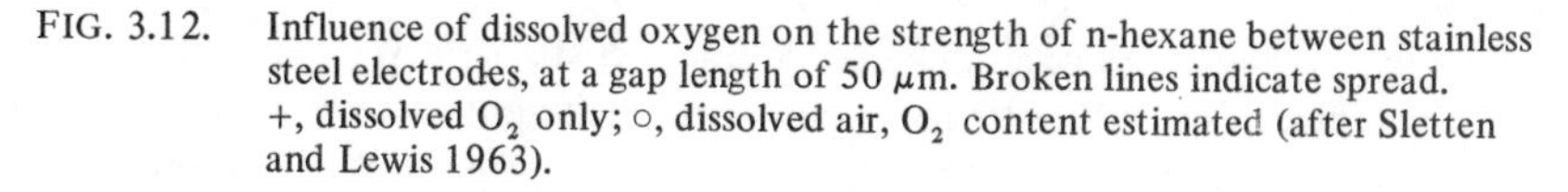

FIG. 3.12. Influence of dissolved oxygen on the strength of n-hexane between stainless steel electrodes, at a gap length of 50 $\mu$m. Broken lines indicate spread. +, dissolved $O_2$ only; ○, dissolved air, $O_2$ content estimated (after Sletten and Lewis 1963).

The minimum amount of dissolved oxygen which can increase the d.c. strength of liquids is not known. It is probably very small since it has been shown by Swan and Lewis (1961) and Gallagher and Lewis (1964 *a*) that only ten parts per million of oxygen in liquid argon can increase its strength. Not oxygen alone, but also nitrogen (Gosling and Tropper 1964), sulphur hexafluoride (Nosseir and Hawley 1966), and other dissolved additives (Angerer 1963, 1965) have been found to alter the d.c. strength of mineral oils. The general pattern has been that small amounts of these gases produce an increase in strength, whereas larger quantities may cause the strength to fall due to the gassing properties of oils (cf. Zaky and Hawley 1973).

In contrast to results obtained for direct voltages, several workers have reported that small traces of dissolved air or oxygen in benzene (Edwards 1951), transformer oil (Watson 1955), n-hexane (Sletten and Lewis 1963), and argon (Gallagher and Lewis 1964 *a*) do not change the mean strength when pulse voltages of microsecond durations are used. It would appear, therefore, that the influence of oxygen is time-dependent and that the most probable mechanism for its effect on strength can be associated with its electro-negative properties. It is well known that oxygen removes electrons from the discharge process in gaseous breakdown by causing electron attachment to form negative oxygen ions with low mobilities, and we have already discussed in sub-section 1.3.1 the process of electron attachment to oxygen which is likely to occur in liquids. Therefore, we shall only discuss here the various ways in which oxygen may contribute to altering the breakdown of liquids. The arguments should apply to any gas possessing an affinity for electrons. Firstly, oxygen may lower the strength by supplying negative ions to establish a space charge at the anode (section 3.4). Secondly, electron capture by oxygen molecules in the liquid will reduce the number of electrons available, which could result in a reduction of collision ionization in the bulk liquid and an increase in strength. Experimental evidence to indicate that oxygen may remove some high energy electrons, capable of ionization, can be seen in the much heavier deposite of wax found in degassed, as compared with air-saturated, hydrocarbons. Thirdly, any positive ions produced in a liquid and normally returning to the cathode may be neutralized, again by attachment to negative oxygen ions. This process may be very important when positively charged particles are present, since they could acquire a layer of negative ions, partly neutralizing their positive charge and reducing their ability to promote emission from the cathode (Sletten and Lewis 1963). These latter processes will increase the strength.

If electron attachment occurs it should be present under pulse voltages also and oxygen should change the pulse strength. As already stated, numerous workers have failed to detect any influence when oxygen concentrations were used, which produced significant changes in the d.c. strength. We can only conclude from these results that the probability of attachment of an electron is small during its journey from cathode to anode. For oxygen in liquid argon the mean-free-path for capture of electrons can be estimated from the measurements of attachment cross-section $Q$ made by Swan (1963). The atomic number density of liquid argon is $2 \times 10^{28}$ m$^{-3}$ and the argon used for breakdown measurements contained approximately $2 \times 10^{-3}$ per cent of oxygen, corresponding to an impurity density, $N$(m$^{-3}$), of $4 \times 10^{23}$ molecules of oxygen. Taking $Q$ for a three-body process to be $5 \times 10^{-22}$ m$^2$ at the highest stress of 11 MV m$^{-1}$ used by Swan (1963), the mean-free-path $\lambda = (NQ)^{-1}$, is $5 \times 10^{-3}$ m. This value is much greater than the electrode spacings which are normally used for breakdown measurements on liquid argon. Therefore, it is not surprising that the mean pulse strength is unaltered by small amounts of dissolved oxygen, but requires the addition of some 10 per cent before an appreciable increase is measured (Gallagher and Lewis 1964 *a*). However, for organic liquids, the situation is less clear. Of these liquids n-hexane is the only one in which a mean-free-path for the capture of electrons has been measured. Devins and Wei (1972) have quoted 0.13 $\mu$m (Table 1.4.2) as the average trapping distance for electrons, presumably to oxygen molecules, in air-saturated n-hexane. It is not possible to calculate accurately the oxygen concentration in the n-hexane used by Sletten and Lewis (1963), but, from the data given in their paper, it would appear to lie between 1 per cent and 10 per cent i.e. $10^3$ and $10^4$ times greater than in argon. The density of n-hexane is 650 kgm$^{-3}$ and, on the dubious assumption that $Q$ is the same for both liquids, we find that $\lambda$ lies in the range from 40 $\mu$m to 4 $\mu$m. These values are somewhat smaller than the electrode spacing of 50 $\mu$m used by Sletten and Lewis (1963) and in view of the uncertainties involved in the estimation of $\lambda$, the smaller value of 4 $\mu$m is in fair agreement with that given by Devins and Wei (1972). With $\lambda = 40$ $\mu$m no difference would be detected in pulse strength between degassed and air-saturated liquids. However, if 4 $\mu$m is a valid estimate for $\lambda$, then a difference would be expected between their mean pulse strengths, but this has not been found. It was suggested by Sletten and Lewis (1963) that the main role of oxygen is not the removal of electrons capable of causing breakdown, but rather the creation of negative ion layers at the anode and around charged particles in the bulk of the liquid. Whereas, under direct stress electron trapping can, over long periods, produce appreciable numbers of negative ions, it is probable that the pulses used were too short to allow the accumulation of similar numbers in the liquid or near the anode. Another possible reason is that, at the higher stresses required for pulse breakdown, the mean electron energy will be greater than for direct voltage tests. Thus, since attachment is a low energy process, its occurrence would be less probable with pulses. Evidence in support of these ideas can be seen in the experiments of Ward and Lewis (1963) who showed that the pulse strength of n-hexane can be changed appreciably by the application of a d.c. stress to the liquid before pulses are applied. This procedure is known as prestressing and is classed as positive or negative according to whether the d.c. stress is of the same or opposite polarity to the pulses subsequently applied. Fig. 3.13 shows that the strength of air-

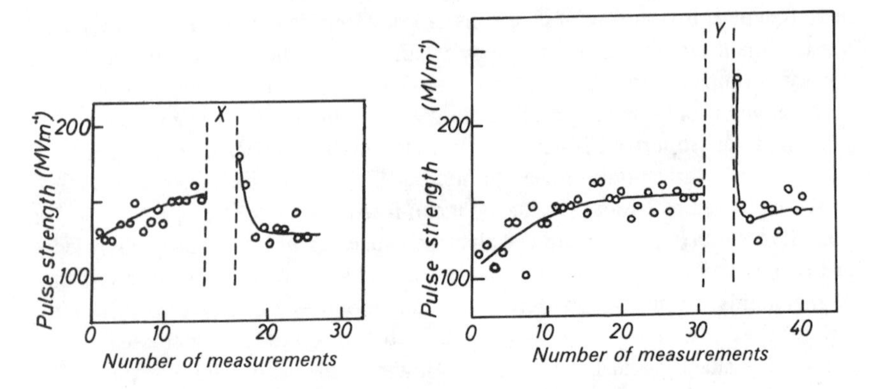


FIG. 3.13. The effect of prestressing on the pulse strength of air-saturated n-hexane. At the points marked X and Y respectively, a positive or negative static stress of 80 MV $m^{-1}$ was applied for 15 minutes before resuming measurements (after Ward and Lewis 1963).

saturated n-hexane can be raised from 150 MV $m^{-1}$ to over 200 MV $m^{-1}$ by using positive or negative prestressing at 80 MV $m^{-1}$. For degassed liquid, the strength was raised or lowered according to whether the static stress was of opposite or the same polarity. From a large series of observations, Ward and Lewis (1963) have concluded that the effects of prestressing are intimately connected with oxygen ion layers and wax deposits on the electrodes. Thus, negative prestressing will produce a negative space charge at the anode, which will reduce its emission when it is used as a pulse cathode. On the basis that emission is important for breakdown, this should increase the strength, as has been observed (Fig. 3.13). The electrode layers set up by prestressing are quite tenacious and even changing the liquid does not seriously disturb them. The results of these prestressing experiments also add considerable weight to the suggestion in section 3.5 that an anode influence on the strength of hydrocarbons is probably present in breakdown measurements because of a strong blocking layer at that electrode.

We have now examined, in some detail, the effects of dissolved gases on the strength of liquids. It is obvious that impurities such as oxygen have produced discrepancies between the results of different investigators and it is important that future work should be carried out in conditions of the best possible chemical purity of the liquid.

## 3.8. Effect of solid impurities

In sub-section 2.3.1, we examined the theoretical and experimental behaviour of particles subjected to an electric stress, and we then considered their contributions to conduction currents in liquids. The pattern in this section will be to scrutinize only the experimental evidence relating to particle effects at breakdown. This should enable us to decide whether the strength obtained in conventional test systems is largely determined by particles, which is currently the opinion of several investigators.

We noted in sub-section 2.3.1 that the high-field region of the gap may contain solid particles even before a measurement is made. After a breakdown, contamination is certainly present in the gap. Each discharge is accompanied by erosion of metal fragments from electrodes, evolution of gas bubbles and, when hydrocarbons are involved, the formation of carbon and wax. As most investigators have used a multi-shot technique (section 3.4) appreciable quantities of discharge products accumulated and particles were undoubtedly present in the gap. Direct evidence for a reduction in the strength by particles has been obtained from experiments on a liquid before and after filtration. For mineral oils many tests have been made using various types and grades of filter (cf. Kok 1961, Crawley and Angerer 1966) and it has been generally found that the strength of the oil increased progressively as the filter pore size was decreased. Salvage (1951) has observed similar results for n-hexane. The mean d.c. strength before filtration was 77 MV $m^{-1}$ but after filtering the liquid to remove particles down to 1 $\mu$m in size the mean strength rose to 88 MV $m^{-1}$. A quite staggering degree of filtration has been obtained in the case of liquid helium by the use of superleak filters. These materials possess pore sizes not much larger than several molecular diameters (Table 3.8.1) and have been used to separate the normal component of liquid helium from its superfluid component (Staas and Severijns 1969). As shown in Fig. 3.14, Goldschvartz and Blaisse (1966) obtained a large increase in the strength of helium after it was purified by superfluid flow through porous Vycor glass. Incidentally, curves A, B, and C of Fig. 3.14 were obtained on samples purified in a similar manner, but the differences between the results serve to illustrate the difficulty in reproducing measurements in this liquid. Despite the use of better techniques more recent values for the strength (Goldschvartz *et al.* 1972) are significantly lower than those of curves *A, B*, and *C* in Fig. 3.14. The dispersion in results has been attributed to changes in the purity of the helium used (Goldschvartz and Blaisse 1966), but it would appear that a less well-controlled experimental factor is involved (Galand 1968). Sletten and Lewis (1963) have noticed that, at breakdown, particles were completel ejected from the high-field region of the gap, but the recovering field after breakdown caused most of them to return. The particle density increased with successive breakdowns but this did not appear to affect the measured strength since its average over some twenty breakdowns was the same. Also, no visual correlation between breakdown and particle behaviour could be established. However, it was observed that when the high-field region between the electrodes was cleared of particles the d.c. strength of oxygenated n-hexane could be as high as 200 MV $m^{-1}$ compared with the usual value of 130 MV $m^{-1}$. Sletten and Lewis (1963) concluded that at d.c. stresses of 130 MV $m^{-1}$ and below, dust

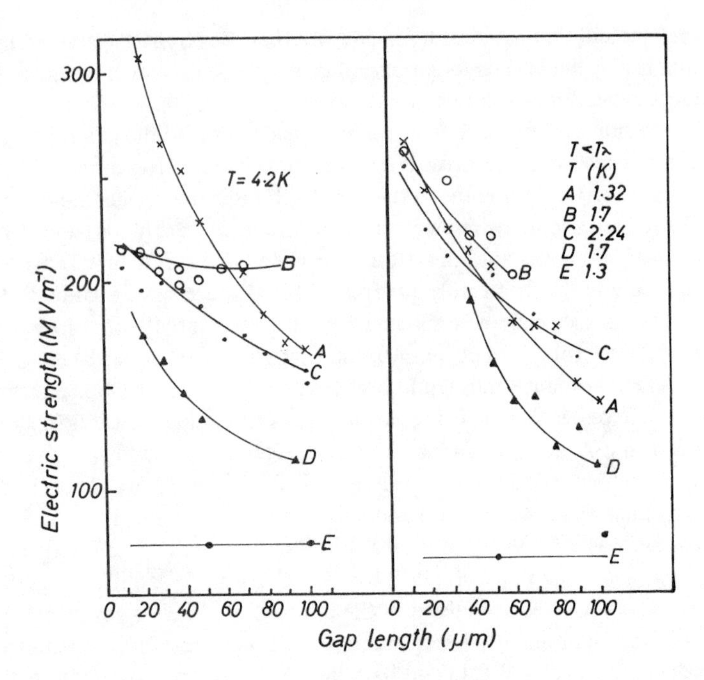


FIG. 3.14. Electric strength of liquid helium at different temperatures as a function of gap length. The measurements were taken over a 6 year period. Curves A, B, and C refer to helium purified by passage through a superleak filter. Curves D and E without purification (after Goldschvartz and Blaisse 1966).

Table 3.8.1.

*Superleaks (after Goldschvartz and Blaisse 1970)*

| Type | Onset† temperature/K | Diameter of pores/Å |
|---|---|---|
| Jeweller's rouge | 1.8 | 100 |
| Vycor glass | 2.0 | 50 |
| Silicon carbide | 1.79 | <70 |
| Wonderstone | 2.1 | 70 |
| Talc-stone | 1.69 | <70 |

† This corresponds to the temperature at which the superfluid component of HeII starts to pass through the material.

particles were largely responsible for the initiation of breakdown in their experiments, and probably, in all previous investigations.

If pulse voltages, of short duration, are used to test liquids it might be expected that the effects of particles can be reduced, if not eliminated, since they would not move within the period of voltage application. However, particles may still encourage breakdown to occur as it has been noticed that particles were thrown from the electrodes after the application of a pre-breakdown pulse (Sletten and Lewis 1963). Once ejected into the liquid they quickly moved to the nearest electrode but, after many breakdowns, their activity ceased and they remained in suspension in the liquid. Using a single-shot technique (section 3.4), 10-microsecond rectangular pulses, and phosphor-bronze electrodes, Edwards (1952) has measured individual values of strength, for a number of liquids, which were 20 to 100 per cent higher than the highest average values reported previously (Table 3.8.2).

Table 3.8.2.
*Electric strengths of polar and non-polar liquids (after Edwards 1952)*

| Liquid | Edward's maximum value/(MV $m^{-1}$) | Highest value previously reported in the literature/(MV $m^{-1}$) |
|---|---|---|
| Carbon tetrachloride | 479 | 248 |
| Chloroform | 253 | 207 |
| Methylene chloride | 236 | – |
| n-Hexane | 246 | 137 |
| Benzene | 183 | 145 |
| Ethyl alcohol | 163 | 127 |

Edwards has attributed these values to the complete absence from the gap of particles which were removed by the use of electrode wipers and by the application of low voltage pulses to the liquid. However, another factor contributing to these high values, which, incidentally, have not been found in more recent work, could arise from the formation of barrier films on the highly-polished electrodes (Sharbaugh and Watson 1962). The latter suggestion is reinforced by our conclusions in section 3.4 that highly polished surfaces are probably covered by a layer of oxide; a surface layer on phosphor-bronze could be substantial. The effects of this layer coupled with a single-shot technique could give rise to higher strengths.

Several investigators have measured the strength of transformer oil which they deliberately contaminated with particles. Darveniza (1969) examined the 60 Hz strength of oil, doped with concentrations of 15 to 500 parts per million by weight of carbon particles, whose diameters ranged from 10 $\mu$m to 50 $\mu$m. Stressing the oil with a slow rate of rise of voltage of $\frac{1}{2}$kV $s^{-1}$ yielded very low strengths which were associated with the presence of a cloud of mobile particles bridging the high-stress region of the gap. With a rate of increase in voltage of 2kV $s^{-1}$ breakdown occurred at much higher voltages which were attributed to the dispersal of these particle bridges by field-induced liquid motion. This suggestion was confirmed by direct experiment as it was found that a forced flow of oil through the gap prevented particle accumulation. A similar explanation has been given for an increase in the a.c. strength of 'pure' fluorocarbon liquids (Boone and Vermeer 1972) and transformer oil (Nelson, Salvage, and Sharpley 1971), under forced circulation, at velocities of several mm $s^{-1}$. Goswami, Angerer, and Ward (1972) have added selenium, silicon, or copper particles to transformer oil and have measured its strength at different gaps and over a range of frequencies from 20 to 1000 Hz, as shown in Fig. 3.15. At any gap length the formation of particle bridges between the electrodes was observed only if the applied stress lay between a threshold and a critical value, which were both dependent on frequency. Bridges, already formed, were destroyed at a gap-clearing stress which was 30 per cent higher than the critical stress but was still less than 50 per cent of the stress at breakdown.

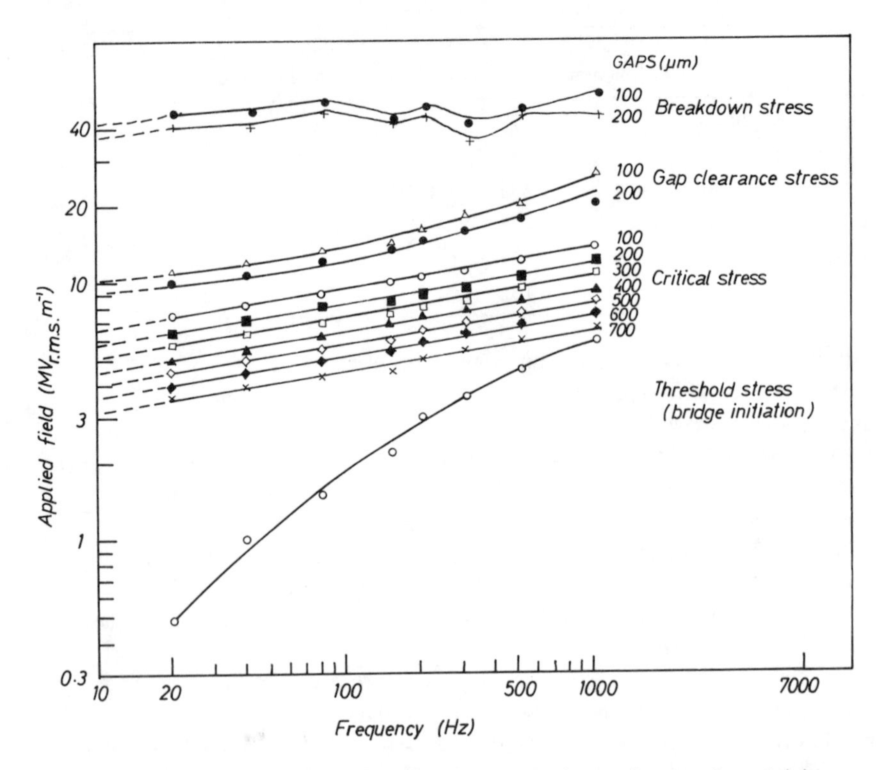


FIG. 3.15. Bridge formation as a function of frequency in transformer oil, containing selenium or silicon particles of sizes ⩽ 45 μm (after Goswami *et al.* 1972).

Kok and Corbey (1956, 1957 *a, b*, 1958) have proposed a theory of breakdown in insulating liquids which is based solely on particle effects. According to these authors the dielectrophoretic force acting on particles, with permittivities greater than that of the liquid, will drive them to the region of maximum stress between the electrodes. Here, the particles can align themselves with the field to form a conducting channel or bridge across the gap which can lead to breakdown via Joule heating of the bridge and the surrounding liquid. Whilst the model may explain the electrical properties of liquids containing large amounts of particles, such as colloid suspensions, it is unlikely to extend correctly to the breakdown of 'pure' dielectric liquids for reasons which have already been stated in section 3.2. Also, particles have been seen to bridge the gap, in n-hexane for example, whilst a discharge occurs in a different region and at a higher voltage (Sletten and Lewis 1963), so that breakdown must involve processes other than those caused only by particles. Nevertheless, particles may, in some cases, be instrumental in the initiation of breakdown.

It has been suggested by Sletten and Lewis (1963) that if the concentration of positively charged particles arriving at the cathode is sufficiently intense, electron streams could occur and initiate breakdown. Particles may also serve as suitable sites to nucleate bubbles and Krasucki (1966) has considered a breakdown to be due to the growth of vapour bubbles which develop in the region of conducting particles (sub-section 3.9.4). An additional effect proposed by Dakin and Hughes (1968) is associated with the generation of a microdischarge as a charged particle leaves or strikes an electrode surface, resulting in a type of 'trigatron' action as in gases. Gzowski, Liwo, and Piltkowska (1968) have used a type of trigatron test cell to study breakdown initiating events in n-hexane. They observed that, when breakdown was triggered by a microdischarge, the stress was some ten times lower than the normal withstand stress of the gap. The results were explained using ideas which were essentially the same as those of Sletten and Lewis (1963) except that positive ions, rather than particles, were believed to play the major role in producing emission from the cathode.

We have now considered, in some detail, the effects of particles on the electric strength of liquids. From the evidence presented it would appear that a reduction in the d.c. and a.c. strength is caused by solid impurities since filtration improves the insulating properties in most cases. Particles may lower the pulse strength but their effects are less obvious and further work is necessary to clarify this point. However, the presence of particles may not always have the catastrophic effect on strength envisaged by some workers; the fact that a long sequence of breakdowns can be carried out without an appreciable fall in mean strength suggests that wax or oxide layers on the electrodes have a greater stabilizing influence on the results than the accumulation of particles in the liquid.

At the end of section 3.7 we concluded that future breakdown measurements should be made on liquids of the best possible chemical purity. It must be obvious now that the physical purity of a liquid is also important. However, the task of removing all insoluble solid impurities, if not impossible, is certainly a most daunting challenge to the experimentalist. Even if this degree of purity is ever achieved it may not be worth the effort as we must remember that even a single breakdown will provide some discharge products! However, the situation is not entirely hopeless. The results of sections 3.7 and 3.8 indicate that, by using pulse voltages, we may be able to minimize the role of gaseous and solid impurities since, under these conditions, they appear to have only a minor influence on mean electric strength.

## 3.9. Liquid effects

Up to now we have discussed most of the parameters other than liquid properties which can influence the strength of a liquid, and by this stage, the reader may wonder if a liquid plays any significant role in a measurement of its breakdown strength. Obviously, experimental values of the strength are determined by the overall electrode–liquid system and are not characteristic of an intrinsic strength for the liquid alone. Because of electrode effects it is doubtful whether this property of a liquid will ever be measured. An intrinsic strength may lie in the region of $10^{10}$ V m$^{-1}$ : field emission experiments on liquid nitrogen by Coelho and Sibillot (1969) have shown that the field at the tip electrode could reach $10^{10}$ V m$^{-1}$ without any sign of pre-breakdown phenomena. Nevertheless, by controlling experimental parameters such as electrode material, gas content, temperature, type of applied stress etc., we can observe changes in strength due to changes in the molecular properties of liquids.

Most of the experimental measurements concerned with liquid effects such as molecular structure (sub-section 3.9.1), temperature (sub-section 3.9.2), and pressure (sub-section 3.9.3) were completed by the early 1960s and, at that time, several theories such as those of Lewis (1956) and of Adamczewski (1957), involving collision ionization, had been developed to explain the results. All of this work has been covered in the comprehensive review articles by Lewis (1959) and by Sharbaugh and Watson (1962). In the interim, new theories based on particle and bubble effects have tended to dominate the breakdown scene and collision ionization has generally been ignored. Notwithstanding the reviews mentioned above, we shall briefly re-consider the earlier experimental and theoretical work as they could not be omitted altogether without leaving our discussion incomplete. Moreover, the recent detection of free electrons in ultra-pure hydrocarbon liquids (sub-section 1.4.2) has again raised the question of ionization being a major factor in breakdown.

### *3.9.1. Influence of molecular structure*

A dependence on molecular structure should be seen in a study of the electric strengths of a homologous series of liquids to which a gradual change of physical, rather than chemical, properties occurs through the series. For example, Salvage (1951) has observed that the d.c. strengths of a series of liquid normal alkanes, n-pentane to n-nonane, increased in a regular manner as the chain length of their constituent molecules was increased. Lewis (1953), also using direct stress, confirmed this finding and later (1958) found a similar pattern with a series of dimethyl siloxanes. Using rectangular pulse voltages, the same trends were found for normal and branched-chain alkanes (Goodwin and MacFadyen 1953, Crowe, Sharbaugh, and Bragg 1954) and for a series of alkyl benzenes (Sharbaugh, Crowe, and Cox 1956). Most of these results have been collated by Lewis (1959) into one figure, which is shown in Fig. 3.16. The strengths are plotted against bulk density, which increases regularly for the alkanes and siloxanes but is approximately constant for the alkyl benzenes. It is obvious that density alone is not the controlling factor because of the large change in strength of the alkyl benzenes, but it is also clear that there is some dependence upon molecular structure.

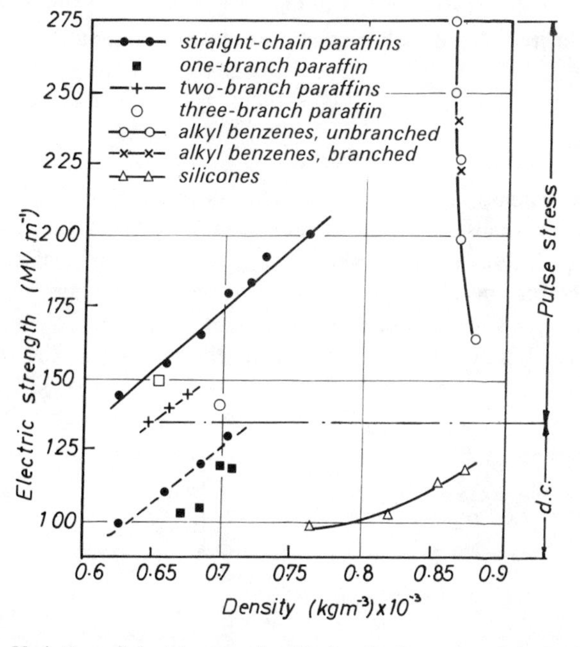


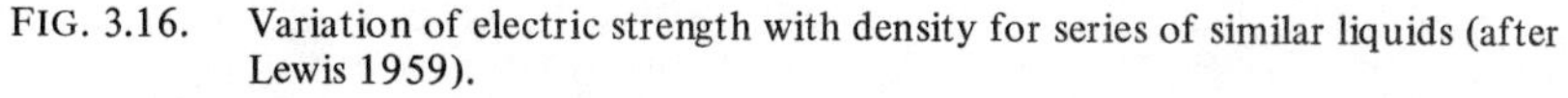

FIG. 3.16. Variation of electric strength with density for series of similar liquids (after Lewis 1959).

Several investigators have attempted to explain the results of Fig. 3.16 on the basis that any influence of structure on strength would manifest itself by controlling the way field-accelerated electrons lose energy to the liquid. The energy distributions for electrons and their collision probabilities have long been known for gases, as are many of the basic processes of conduction and breakdown in solids, but because of our present limited knowledge of the liquid state a detailed understanding of the energy-loss mechanisms of electrons in hydrocarbon liquids is not available (see sub-section 1.4.2). However, having assessed the major collision processes that are likely to occur between electrons and molecules in a liquid, Lewis (1956) adopted a suggestion of von Hippel (1937) that a vibrational collision provides the most efficient mechanism of energy transfer. A hypothetical energy-loss diagram for vibrational collisions is shown in Fig. 3.17.a. The energy gained by an electron from an applied field $E$, below the breakdown value, is represented by a horizontal line $Ee\lambda_m$ where $\lambda_m$ is the minimum mean-free-path corresponding to the maximum cross-section for this type of collision. The distribution of electron energies for the field $E$ is shown in Fig. 3.17.b. At a sufficiently high field, an electron will acquire energy greater than that corresponding to the peak in Fig. 3.17.a when its velocity is so high that the molecular vibrations can no longer respond. Under these conditions, continuous electron acceleration is possible, the distribution in Fig. 3.17.b is shifted to higher energies and ionizing collisions can occur, leading to breakdown. Therefore, the energy balance criterion at breakdown can be expressed as:

$$e E \lambda_m = C \mathrm{h}\nu. \tag{3.6}$$

The quantity $\mathrm{h}\nu$ corresponds to the quantum of energy lost by an electron during a vibrational collision. The constant $C$ takes account of any changes in experimental conditions which are likely to alter the degree of ionization required for a current instability leading to breakdown. For example, the use of pulse or d.c. voltages and the presence of dissolved gases or surface layers on electrodes are some factors which would change the value of $C$. The subsequent

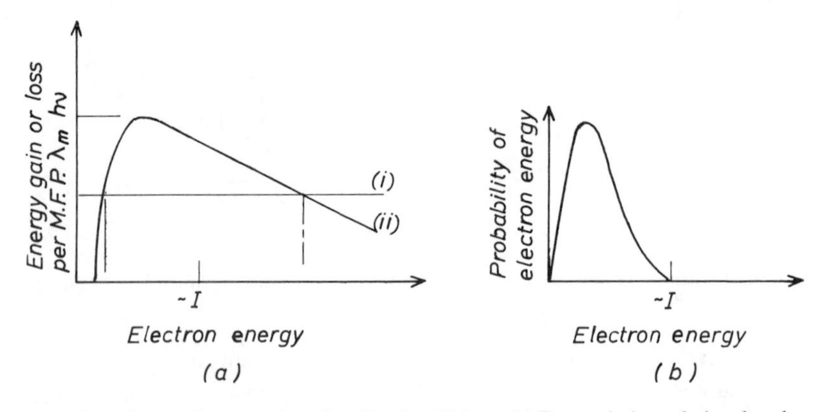


FIG. 3.17. Energy diagrams for vibrational collisions. (a) Energy balance below breakdown stress, (*i*) energy gain from field E, (*ii*) energy loss in vibrational collisions of frequency $\nu$; (b) electron energy distribution function corresponding to (a), $I$ is the ionization energy (after Lewis 1956).

stages of breakdown, once a current instability is reached, are not considered. However, after a build-up of current via charge multiplication and electron avalanche formation, breakdown is thought to progress through the nucleation of bubbles or the formation of streamers, either of which can rapidly lead to a conducting channel propagating across the gap.

In the application of eqn (3.6) to liquid paraffins, Lewis associated the quantity $\nu$ with the stretching vibrational modes of the carbon-hydrogen bonds. These modes are located in the i.r. at a frequency in the region of $10^{14}$ Hz and the quantum of energy involved is approximately 0.37 eV. As we have seen in section 3.7 the mean-free-path $\lambda_m$ can be related to $N$, the density of molecules, and their scattering cross-section $Q$, by $(\lambda_m)^{-1} = NQ$. In addition each molecule was assumed to provide independent collision centres characteristic of its particular groups; i.e. CH, $CH_2$, $CH_3$. Thus $(\lambda_m)^{-1}$ may be written as:

$$(\lambda_m)^{-1} = N \Sigma\, n_i Q_i \quad , \tag{3.7}$$

where $n_i$ represents the number of groups of type $i$ present whose cross-section is $Q_i$. Combining eqns (3.6) and (3.7) the electric strength is given by:

$$E = k N (n_1 Q_1 + n_2 Q_2 + n_3 Q_3) \quad , \tag{3.8}$$

where $k$ is a new constant including $C$, $e$, and $h\nu$, and the suffixes refer to the CH, $CH_2$, and $CH_3$ groups, respectively. For example, the n-alkanes can be described by the general formula $CH_3 \cdot [CH_2]_{n-2} \cdot CH_3$, for which $n_1 = 0$, $n_2 = n-2$ and $n_3 = 2$ in eqn (3.8).

The model may be applied to many different forms of bonding and structure with the only reservation that $C$ in eqn (3.6) should remain reasonably constant. Thus, Lewis (1956) has calculated the relative effective cross-sections of the CH, $CH_2$, $CH_3$ groups in the branched and normal alkanes and has obtained good agreement with these different liquids. When applied to the dimethyl siloxanes (Lewis 1958), the theory gave excellent agreement with the cross-section for the $CH_3$ groups deduced from the alkanes, which was considered as strong evidence for the validity of the breakdown model. However, cross-sections for the $CH_3$ groups in the alkyl benzenes were about twice as large as those obtained from the alkanes. This discrepancy was attributed to the polar nature of the alkyl benzenes and to the degree of ordering of the liquid phase. There are two postulates in Lewis's theory which are not supported by the experimental results discussed in sub-section 1.4.1. Thus, the values for $\mu_e$ and $bd$ in Table 1.4.1 indicate that the molecular cross-section for electron interaction with hydrocarbon liquids is not simply an additive property of the type of groups present in the molecule; the cross-section depends also on the spatial arrangement of the groups. Moreover, the major barrier to electron acceleration would not appear to involve vibrational excitations, as suggested by Lewis, but, rather, Raman-type rotational processes.

Adamczewski (1957, 1969) has also proposed a model for hydrocarbon liquids which is similar to that of Lewis, but with the C–C bond vibrations acting as the major barrier to electron acceleration. Furthermore, the total cross-section for excitation of molecular vibrations was estimated on the basis that it was equal to the longitudinal geometrical cross-section of the molecule. Thus, the collision cross-section of an n-alkane molecule was given as:

$$Q = dl(n-1) \quad , \tag{3.9}$$

where $d$ is the mean distance between molecular axes, $l$ is the C–C bond length projected on the axis, and $n$ is the number of carbon atoms. Using eqn (3.9) in eqns (3.6) and (3.7) with $C = 1$, the electric strength was expressed as:

$$E = \mathbf{h}\nu dl(n-1)N. \tag{3.10}$$

By inserting established experimental data for the various physical quantities in eqn (3.10), Adamczewski has made calculations of the absolute electric strength for a large number of aliphatic and aromatic hydrocarbons and has obtained excellent agreement with the experimental results of Sharbaugh *et al.* (1956). The estimated values for the n-alkanes are shown in Table 3.9.1.

Table 3.9.1.
*Electric strengths of n-alkanes (after Adamczewski 1957, 1969)*

| Liquid | $E$(MV m$^{-1}$) Theoretical from eqn (3.10) | Experimental | Standard deviation per cent |
|---|---|---|---|
| n-Pentane | 144 | 144 | 3.2 |
| n-Hexane | 157 | 156 | 2.8 |
| n-Heptane | 168 | 166 | 2.7 |
| n-Octane | 177 | 179 | 4.0 |
| n-Nonane | 185 | 184 | 5.2 |
| n-Decane | 191 | 192 | 4.2 |

The agreement in Table 3.9.1 is fortuitous for several reasons. In the first instance, the energy quantum assigned to the C–C bond vibration was taken as 0.114 eV but the use of this unique value is unrealistic. The values of wavelength usually assigned to the stretching frequencies are only quoted to within $\pm 10^4$ m$^{-1}$ (Herzberg 1945), corresponding to an accuracy in their energies of $\pm 0.012$ eV. Keeping the other terms in eqn (3.10) constant and inserting these limits for $h\nu$ would change the estimated values in Table 3.9.1 by approximately $\pm 10$ per cent. Secondly, estimates of the collision cross-section, based on eqn (3.9), would probably increase if allowance were made for a random orientation of molecules due to thermal fluctuations. In fact, the semi-empirical values quoted in Table 1.4.3 show that the scattering cross-sections have approximately double the area estimated on the basis of Adamczewski's geometrical model. Direct comparison is possible only for two alkanes. The geometrical cross-section is 24.1 Å$^2$ for n-pentane, and 30.13 Å$^2$ for n-hexane, compared with 51.5Å$^2$ and 59.2Å$^2$, respectively, in Table 1.4.3. Finally, eqn (3.6) was used with $C = 1$, which neglects the effect of different experimental variables on the electric strength. On the other hand, the results in Table 1.4.3 indicate that the collision cross-section increases with the number of carbon atoms per molecule. Hence, if the critical mechanism of breakdown involves electron interaction with the C–C bonds, as suggested by Adamczewski, the electric strengths of normal liquid paraffins should increase with an increase of chain length. For the n-alkanes in Table 3.9.1 the predicted order is correct.

Since about 1960 the foregoing *electronic theories of breakdown*, as they are usually called, have been largely rejected on the following grounds:

(*i*) the mean-free-path of electrons in a liquid would be too small to enable them to acquire an ionization energy which, for hydrocarbons, is about 10 eV. With applied fields of 100 MV m$^{-1}$ a mean-free-path of $10^{-7}$ m would be needed;

(*ii*) there was no direct experimental evidence for an $\alpha$-process, and;

(*iii*) they did not predict any influence of hydrostatic pressure on electric strength (sub-section 3.9.3).

With reference to (*i*) and (*ii*), it has been claimed that the existence of energetic electrons is confirmed by the observations of visible light emission from insulating liquids subjected to high electrical stresses (Dakin and Berg 1959, Darveniza and Tropper 1961). These observations of electroluminescence were seen as direct evidence that electrons could reach energies of 2 to 3 eV in liquids and that some of those in the high energy tail (Fig. 3.17.b) of an electron swarm could reach the ionization energy. The presence of energetic electrons may also be inferred from the fact that wax-like deposits are found on electrodes after high-field conduction studies in hydrocarbon liquids (sub-section 2.3.2). Since it requires about 4 eV to dissociate a C–H bond, the presence of 10 eV electrons does not seem improbable. Moreover, the high mobilities in ultra-pure hydrocarbons provides new evidence to indicate that electron mean-free-paths could be significantly larger than hitherto assumed possible. On the other hand, the emission of pre-breakdown light is thought to arise from a type of gas discharge in a bubble in the liquid (Smith and

Calderwood 1968), whilst electroluminescence has been attributed to partial breakdowns caused by the arrival of charged dust particles at the cathode (Gzowski *et al.* 1966; Gzowski 1966). The latter conclusion has been rejected by Zaky and Hawley (1973) who have analysed, in some detail, the reports of light emission in liquids. At the beginning of this section we implied that the electric strength is dominated by conditions at the electrode–liquid interfaces. Leaving aside electrode effects, a significant part of the influence of molecular structure might arise through excitation collisions with epithermal electrons (see subsection 1.4.1) emitted from the cathode. If this idea is correct, then, some correlation between electric strength and molecular structure should exist, as was observed by Lewis, and by Adamczewski. Thus, it may be important to incorporate some aspects of the *electronic theories* into new models of breakdown.

### *3.9.2. Effect of temperature*

Salvage (1951), Edwards (1951), Lewis (1953), Goodwin and MacFadyen (1953), Kao and Higham (1961), and Brignell and House (1965) were the only investigators to make detailed measurements of the variation of strength with temperature. However, some of these measurements refer to single tests on a sample of liquid, which must place some doubt on their true significance. Nevertheless, all results for both non-polar and polar organic liquids, under direct and pulse voltages, appear to have similar characteristics, with the sole exception of the recent work by Brignell and House (1965) on n-hexane. Therefore, we shall confine our attention to measurements on this liquid.

There is widespread agreement amongst all results that the electric strength falls off rapidly at temperatures near the boiling point, but there is some confusion over the trends occurring near room temperatures and lower. Fig. 3.18 contains measurements of the strength of n-hexane, covering the widest range of temperatures studied. In Fig. 3.18 curve *a* and curve *d* were obtained using pulse voltages with fresh electrodes and liquids while curve *b* refers to direct stress tests on one pair of electrodes and one sample of liquid. Despite these very different experimental conditions we can detect a general pattern of a gradual change in strength near room temperature, with a considerable increase at lower temperatures. Kao and Higham (1961) qualitatively explained their findings on the basis of easier vaporization in the liquid as its temperature is increased, whereas Lewis (1957) concluded that liquid structure, as distinct from molecular structure and density, has an important influence on the strength. As can be seen from curve *b* in Fig. 3.18 there are marked changes in strength near −20 °C and between 20 °C and 50 °C. These transition temperatures correspond closely with the temperatures at which the degree of local order of n-paraffin liquids undergoes considerable change due to the onset of vibrations and rotations of the molecules. As the order is destroyed so the strengths of the liquids fall. The similarity between curves *a, b,* and *d* of Fig. 3.18 would make this argument

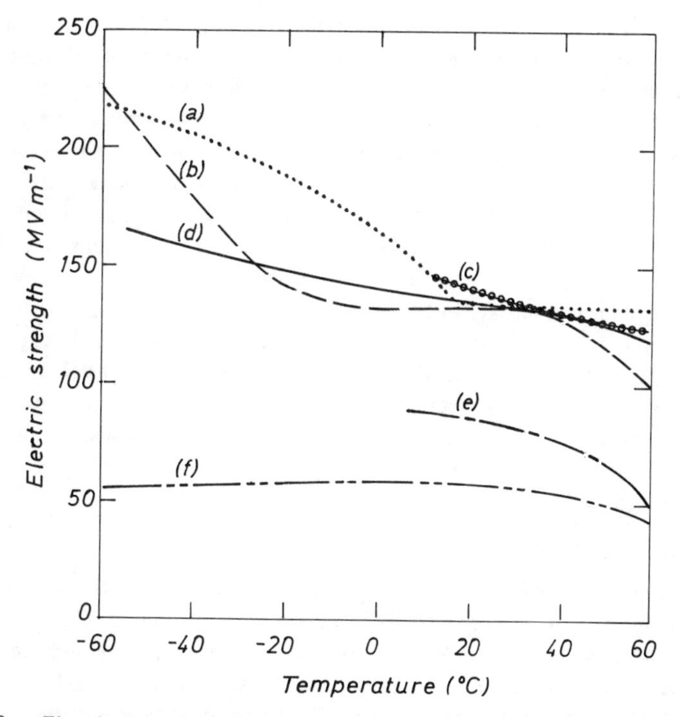


FIG. 3.18. Electric strength of n-hexane as a function of temperature. (a) Edwards (1951); (b) Lewis (1953); (c) Goodwin and MacFadyen (1953); (d) Kao and Higham (1961); (e) Salvage (1951); (f) Brignell and House (1965).

plausible were it not for the direct stress results of curve *f* in Fig. 3.18. These measurements were taken on twelve samples of degassed liquid and they indicate a constant strength over the same range of temperatures. Brignell and House (1965) have suggested that a temperature effect may have figured in earlier measurements through a change in the mobility of an oxygen ion in the liquid but this explanation seems unlikely. Oxygen does not appear to alter the mean pulse strength (section 3.7), whilst any temperature-induced change in ion mobility is probably not large enough to be important in tests with direct stress.

From the limited results available, we cannot arrive at a satisfactory explanation of the influence of temperature on strength. The n-hexane samples used by Lewis (1953) and by Kao and Higham (1961) most probably contained some oxygen and gave results in marked contrast to those taken on degassed n-hexane (Brignell and House 1965). It would appear, therefore, that temperature has some effect on gas layers, notably oxygen, at the electrode–liquid interfaces but the matter needs further clarification.

The temperature of cryogenic liquids can be altered by controlling their vapour pressures, but most investigators have restricted their measurements to temperatures very close to the boiling point, except in the case of liquid helium.

Goldschvartz *et al.* (1972) have concluded, after many years of study on helium, that, above the λ-point, its mean strength is almost independent of temperature, in agreement with the results of Blank and Edwards (1960) and Gerhold (1972). Below the λ-point they have observed a small decrease in strength, whereas Blank and Edwards found an appreciable drop in strength, as already shown in Fig. 3.7. For the other liquefied gases a few isolated measurements indicate that the strength increases at temperatures below their boiling points, as is the case with simple organic liquids (Fig. 3.18). However, unlike the latter, density does not appear to influence greatly the magnitude of the electric strength of cryogenic liquids; liquid argon has a greater density but lower strength than either oxygen or nitrogen while helium and hydrogen have somewhat comparable strengths although their densities are nearly ten times smaller. On the other hand, liquid structures may be very important as there is a high degree of local order in several of the liquefied gases (Henshaw *et al.* 1953). A detailed study of breakdown in these liquids over a wide temperature range could determine the importance of liquid structure.

*3.9.3. Effect of pressure*

Kao and Higham (1961) are the only authors who have studied the effects of hydrostatic pressure on the strength of a large range of liquids. The compounds investigated included carbon tetrachloride, n-hexane, n-heptane, n-decane, benzene, toluene, chlorobenzene, ethyl alcohol, and methyl alcohol. A representative set of results, in this case for partially degassed n-hexane, is shown in Fig. 3.19, where the strength is seen to increase with an increase in pressure, for rectangular pulse durations ranging from 1 $\mu$s to 1 ms. These results refer to a single breakdown with each liquid sample and pair of electrodes, so that the usual conditioning process was purposely avoided. Taken in conjunction with their measurements on temperature effects, a pressure dependence was seen by Kao and Higham (1961) as further evidence that breakdown is initiated in a bubble of gas or vapour. In favour of a bubble theory, Kao and McMath (1970) have argued that a certain time is required for bubbles to form and to elongate under electric stress and, consequently, a pressure dependence would disappear if a liquid was tested with very fast-rising voltages. The arguments appear to be verified by their experimental results, as shown in Fig. 3.20. Using ramp-function voltages of various rates of rise they have measured the strengths of n-hexane and transformer oil under gauge pressures of 0 and 15 bar. In an attempt to accentuate the pressure effect a large spacing of 910 $\mu$m was used, and the electrodes were also well conditioned before testing. In both liquids a time-dependent pressure effect was observed, which diminished with increasing rate of rise of applied field and which practically disappeared at rates of rise faster than 400 MV m$^{-1}$ per $\mu$s. Above these fast rates of rise a different breakdown mechanism was postulated, involving multiple electron avalanches in the liquid but this point was not elaborated. Sletten and Lewis (1963) have reported that

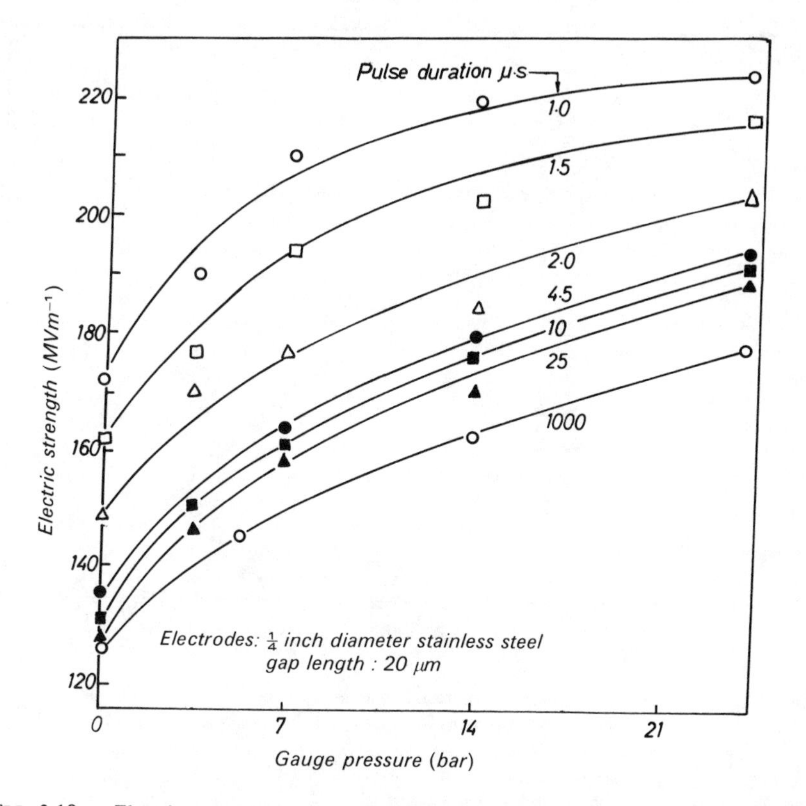


FIG. 3.19. Electric strength as a function of pressure for degassed n-hexane with a range of pulse durations (after Kao and Higham 1961).

the strength of degassed n-hexane under direct stress falls as the pressure is reduced below atmospheric, but that any pressure dependence was very sensitive to the experimental techniques used. They have suggested that gas attached to the electrodes may be very important, presumably in the nucleation of bubbles. The complete removal of absorbed gas can only be achieved by vigorous outgassing methods (Low 1958) and it is most unlikely that electrode surfaces have ever been truly degassed before breakdown measurements. Hydrostatic pressures could alter the gas in equilibrium at electrode surfaces which in turn could change the emission at the cathode or even the acceptance of charge at the anode. However, the persistent pressure dependence with a point cathode caused Kao and Higham (1961) to dismiss the notion that a bubble is formed from absorbed gases but the possibility that it might be formed by gas attached to the plane anode was not considered. Furthermore, their single-shot procedure was designed to avoid a conditioning process whereas detailed work on transformer oil has shown that conditioning itself is dependent on pressure and probably connected with adsorbed air on the electrodes (Tropper 1961).

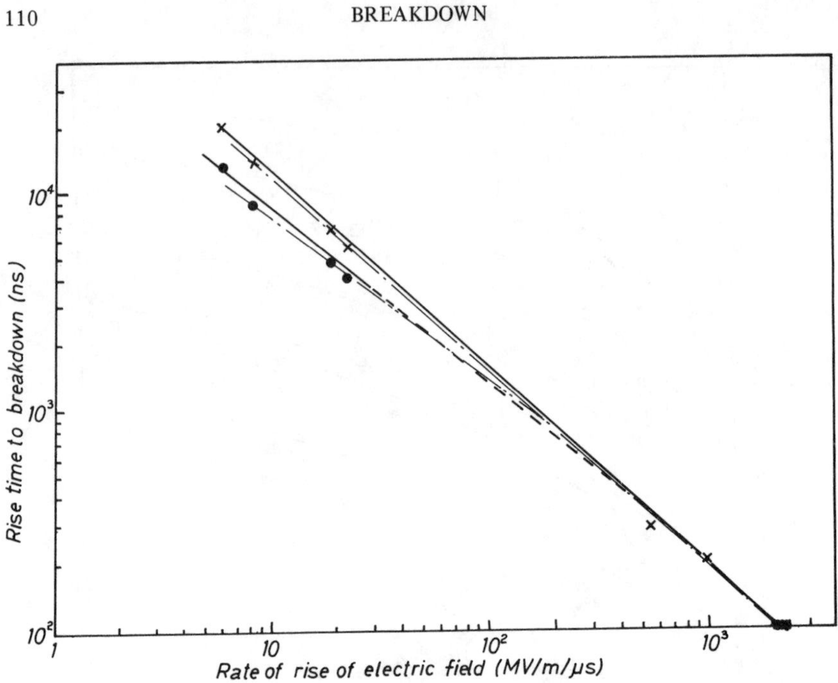


FIG. 3.20. Breakdown volt-time characteristics of n-hexane (—) and transformer oil (– – – – –). Gauge pressures: •, 0 bar; ×, 13.5 bar (after Kao and McMath 1970).

It is also difficult to assess the role of dissolved gases in the nucleation of bubbles. Sletten and Lewis (1963) did not find any difference in strength due to large amounts of dissolved hydrogen, nitrogen, or carbon dioxide whilst the pressure dependence of the strength was the same for air-saturated and for partially degassed liquids (Kao and Higham 1961). Contrary to these findings, the strength (Tropper 1961) and gassing properties (Basseches and Barnes 1958) of transformer oil were markedly influenced by pressure and air content; effects which were ascribed to bubble formation near suspended dust particles and at electrodes. This pattern of conflicting evidence, so common to the field of liquid dielectrics, is probably caused by different experimental procedures. Lewis (1959) has concluded that whenever a liquid contains dissolved gas some pressure effect will be observed, the extent of which will depend on the experimental technique used. The effect may be altered, or sometimes removed, by degassing the electrodes, the liquid, or both.

Since the measurements of Kao and Higham (1961) there has been a tremendous accumulation of evidence from investigations of pre-breakdown phenomena to support the idea that a discharge in organic liquids is preceded by bubble formation. We shall only briefly summarize this work as there is not sufficient space in a book of this size to deal adequately with all of the results. Point–plane or point–point electrode geometries together with a wide variety of sophisticated experimental techniques have been used in these studies. Several investigators (Hakim and Higham 1961; Farazmand 1961; Chadband and Wright 1965; Thomas 1973 *b*; Allan and Hizal 1974) have used a Schlieren optical system to detect changes in the refractive index of electrically-stressed liquids; changes which were attributed to the growth of a weak conducting plasma from the point cathode. In an effort to separate a breakdown initiating event from the subsequent arc formation, Palmer and House (1972 *a*) have taken exceptionally fast streak photographs of the propagation of a spark in n-hexane.

High-speed cinematography has been used to photograph a spray of bubbles emanating from a very fine point in mineral oil (Krasucki, Church, and Garton 1962) and in n-hexane (Singh, Chadband, Smith, and Calderwood 1972; Chadband and Calderwood 1972). For n-hexane the onset voltage for the spray was only dependent on pressure when the point in a point–plane system was positive (Mirza *et al.* 1970, 1972). A bubble chamber has been used to observe pre-breakdown phenomena in superheated liquids by Murooka, Nagao, and Toriyama (1961, 1964). At values of applied stress in the region of 30 MV $m^{-1}$, bubbles were believed to originate from adsorbed gas on the electrodes but at stresses of 100 MV $m^{-1}$ the primary cause was associated with cathode emission. Observations by means of an image-intensifier (Smith and Calderwood 1968) and light-scattering techniques (Singh, Smith, and Calderwood 1972) have led to estimates of bubble radii of 0.7 $\mu$m near a negative point and 1 $\mu$m near a positive point. On the other hand, Chadband, Coelho, and Debeau (1971), using a shadowgraph technique, have reported that corona-like discharges at a point cathode in n-hexane were much thicker than those at a point anode, and resembled the discharges obtained in transformer oil by Toriyama, Sato, and Mitsui (1964), who used a 'Lichtenberg figures' technique. Complete agreement between results is unlikely because of the different tip radii and experimental techniques that were used. It should also be noted that only Smith and Calderwood (1968), Palmer and House (1972 *a*), and Thomas (1973 *b*) used techniques with temporal resolutions as low as a few nanoseconds. One point to emerge from this diverse examination of pre-breakdown phenomena is that localized power densities at finely-tipped negative or positive points can produce regions which are weaker electrically than the rest of the liquid. It was generally believed that these regions exist as bubbles, but, in the most recent work of Allan and Hizal (1974) on transformer oil, experimental evidence is presented which suggests that these regions consist of liquid at a higher temperature than the surrounding bulk liquid. Whatever the nature of these regions, ultimately a spark discharge can follow by growth and extension of one of them across the liquid, encouraged either by electron ionization in the liquid or at the 'bubble'-liquid interface. The work on pre-breakdown phenomena has also shown that with non-uniform fields breakdown may even be propagated from a point anode.

The latter observation could account for the hitherto unexplained reduction in pulse strength when the anode surface was roughened in the sphere–sphere measurements of Sharbaugh *et al.* (1956).

With special emphasis on possible engineering applications, the influence of pressure on the strengths of cryogenic liquids has been studied in recent years. By increasing the absolute pressure from 1 to 3 bar, to prevent boiling of the liquid, Jefferies and Mathes (1970) have measured a twofold increase in the 60 Hz breakdown voltage of hydrogen and nitrogen between concentric cylindrical electrodes at a spacing of 6.35 mm. At higher pressures, up to 7 bar, the strengths were virtually constant at about 18 MV $m^{-1}$, which is very much lower than the values usually measured between spherical electrodes at small spacings (Table 3.5.1). The larger electrodes and gap would be expected to yield a lower strength but other factors, especially liquid purity, were probably responsible. Burnier, Moreau, and Lehmann (1970) have observed a slow, linear increase in the 50 Hz breakdown voltage of hydrogen and nitrogen for absolute pressures up to 5 bar. Similar results have been obtained for helium at pressures above the critical value by Fallou, Galand, Bobo, and Dubois (1969) and more recently by Gerhold (1972) for the d.c. strength of subcooled helium. Jefferies and Mathes (1970) have tentatively suggested that their results are reasonably consistent with a 'bubble' theory of breakdown (Krasucki 1966), whilst Gerhold (1972) has attempted a quantitative explanation of his pressure dependence, on lines which are broadly similar to the thermal mechanism of Watson and Sharbaugh (1960).

### *3.9.4. Theories to explain effect of pressure*

In sub-section 3.9.1, we discussed two models of breakdown based on electronic excitation of intra-molecular vibrations and collision ionization in liquids. Whilst these approaches can account for some of the experimental results they cannot explain the fact that the strength of a liquid should depend on hydrostatic pressure. Here, we shall examine the various theories which have been formulated to account, especially, for the effect of pressure, but which have also been extended to explain other experimental observations.

Applied hydrostatic pressures up to 25 bar, the highest used by Kao and Higham (1961), will not cause any significant change in density or electron motion in a nearly incompressible liquid. Therefore, the influence of pressure on strength has been interpreted by many investigators to indicate that the critical stage of the breakdown process is linked with a change of phase, involving the formation of a bubble of gas or vapour. Once a bubble is formed, it is assumed to be elongated in the direction of the applied field by electrostatic forces, and ionization, followed by breakdown, occurs in the gas or vapour when the bubble has attained a certain size. It is appropriate to remark here that one of the major differences between *electronic* and *bubble* theories of breakdown lies in the importance attached to the temporal development of events which precede a spark. In the former case, it is assumed that ionization and current growth begins first in the liquid whereas, in the latter case, these processes begin to occur in the gaseous phase after the nucleation of a bubble. The distinction between the theories may appear somewhat naive but the order of events is vital to the formation of any criterion for breakdown.

Watson and Sharbaugh (1960) have advanced a thermal mechanism for breakdown in which a vapour bubble is generated in the liquid by the injection of large power densities, in the region of $10^{13}$ watts per $m^3$, from asperities on the cathode. By this process local vaporization can occur in only a few microseconds. A criterion for breakdown was developed in terms of the heat $H$ needed to vaporize a liquid from its ambient temperature $T_a$. Thus, for unit mass of liquid, $H$ can be written as

$$H = C_p\,(T_b - T_a) + l_v, \qquad (3.11)$$

where $C_p$ is the average specific heat at constant pressure, $l_v$ is the latent heat of vaporization, and $T_b$ is the boiling temperature. $H$ was related to the applied field $E$ by assuming that the current from a strongly-emitting asperity was space-charge-limited and of the form $I = A\ V^n$, with $n$ in the range of 1.5 to 2. Therefore, for an applied pulse voltage $V$, of duration $\tau$, and with $E$ proportional to $V$, the local energy input has the form

$$H = A\ E^n \tau, \qquad (3.12)$$

where $A$ is an unknown factor of proportionality.
Combining eqns (3.11) and (3.12) we get the breakdown equation

$$AE^n\tau = C_p\,(T_b - T_a) + l_v. \qquad (3.13)$$

The thermal model exhibits a marked pressure dependence of breakdown strength. As pressure on a liquid is increased its boiling temperature $T_b$ increases in a known fashion, and eqn (3.13) was used to compare the calculated strengths with the experimental results, at constant pulse width, of Kao and Higham (1961). In most cases $n = (\frac{3}{2})$ in eqn (3.13) gave the best agreement. An identical value for $n$ was used to fit experimental results to the temperature dependence of strength. The effect of molecular structure can also be incorporated into a thermal mechanism since, within a given homologous series, the thermal properties $C_p$ and $l_v$ increase with chain length. For the normal alkanes a three-halves power law in eqn (3.13) described the variation of breakdown strength but the predicted values for the branched alkanes were too high (Sharbaugh and Watson 1963). As a further test of their model, these authors measured the pulse strength of n-hexane over a range of temperatures and pressures near the critical point in both the liquid and the vapour phases (Watson and Sharbaugh 1963). It was postulated that, at normal room temperatures, breakdown is preceded by the formation of a low strength, low density, vapour bubble and, consequently, it would be expected that liquid and vapour would have the same strength at the critical point. Only at this point are the two phases at the same temperature and pressure. Fig. 3.21 shows that there is no discontinuity in strength in going from the vapour to the liquid state. However, since ionization is necessary to initiate breakdown in a gas the continuity in strength at the critical point need not be taken as proof that electron multiplication in a liquid only occurs after bubble formation. Also during the measurements near the critical point, many bubbles were seen in the liquid which, surprisingly, did not appear to lower its strength and this was explained on the basis that at high temperatures and pressures vapour bubbles have a high density and, therefore, a high electric strength.

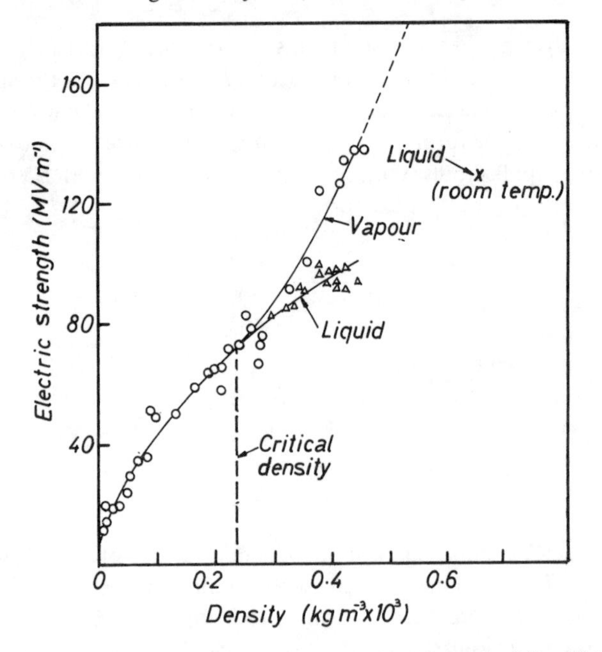


FIG. 3.21. Electric strength of n-hexane as a function of density (after Watson and Sharbaugh 1963).

A thermal mechanism can predict many of the trends in experimental results obtained under pulse conditions. It is also supported by the direct photographic evidence of low density regions in the vicinity of sharp electrodes (sub-section 3.9.3). However, some relationship between strength and thermal properties is to be expected if only on the basis that a discharge always results in the vaporization of a quantity of liquid. This fact is reflected in the simple mathematical treatment presented by Watson and Sharbaugh (1960), which only involves a steady-state equation to describe a transient heat-flow problem. The thermodynamics of breakdown are characterized by irreversible processes and a detailed theoretical analysis of the situation must await more experimental information about thermal mechanisms in the liquid and at asperities on the electrodes.

Krasucki (1966) has suggested that the thermal model of breakdown only refers to specific experimental conditions. In its stead, he has postulated a more general mechanism whereby vaporization will occur in a liquid wherever a point of zero pressure is developed and a vacuous cavity begins to form. Electron bombardment of the walls of the cavity will sustain its growth, eventually leading to a breakdown. Solid impurity particles in the liquid, but more particularly at electrode–liquid interfaces, were considered as suitable sites for the nucleation of cavities. The theory was supported by a study of pulse and direct voltage breakdown in hexachlorodiphenyl, which has an exceptionally large viscosity of $\sim 10^5$ Pa s at room temperature. Consequently viscous forces can retard bubble growth to such an extent that a 'slow-motion' picture of breakdown can be photographed using a conventional cine-camera. With a uniform field geometry at a stress above 100 MV $m^{-1}$ breakdown was seen to occur through the formation and growth in the liquid of dark regions which were preferentially initiated at particles but, in the absence of visible particles, always formed at electrode surfaces. Breakdown of liquids with low viscosities was believed to follow a similar pattern. Photographic evidence, reinforced by theoretical estimates, of their growth and decay led Krasucki to conclude that these dark regions consisted of bubbles of vapour.

To quantify his theory Krasucki assumed that in the vicinity of a particle an enhanced electric field will generate an electromechanical pressure $P_{em}$ tending to lift the liquid off the particle surface against the opposing hydrostatic ($P_h$) and surface tension ($P_{st}$) pressures. The critical condition for zero pressure and cavity formation is then given by

$$P_{em} = P_h + P_{st}. \tag{3.14}$$

Using the appropriate expressions in eqn (3.14) the breakdown strength $E$ may be derived from

$$\left(\frac{\epsilon}{2}\right)(mE)^2 = P_h + \frac{2\sigma}{R}, \tag{3.15}$$

where $\epsilon$ and $\sigma$ are the absolute permittivity ($Fm^{-1}$) and surface tension ($Nm^{-1}$) of the liquid, and $m$ is the field enhancement at the tip of the particle whose radius is $R$ (m). In eqn (3.15) Krasucki put $m = 4.2$ and therefore the breakdown strength is derived as:

$$E = 3.37 \times 10^{-1} \left[ \frac{1}{\epsilon} \left( P_h + \frac{2\sigma}{R} \right) \right]^{\frac{1}{2}} \text{Vm}^{-1}. \tag{3.16}$$

Using eqn (3.16) with particle radii of 100 Å and 250 Å, Krasucki showed that the pressure variation of pulse strength measured by Kao and Higham (1961) is contained within the theoretical estimates, as shown in Fig. 3.22. The particular values of radii were chosen because particles of a similar size were produced by spark erosion of electrodes in liquid argon (Bucklow and Drain 1964). Changes in the surface tension term in eqn (3.16) were also used to predict correct trends for the variation of strength with temperature and with chain length of some n-alkanes.

Although the agreement between theory and experiment is good, the model cannot explain several important experimental observations such as the influence on electric strength of electrode material (section 3.5), electrode separation (section 3.6) and dissolved gases (section 3.7). We have seen in section 3.8 that

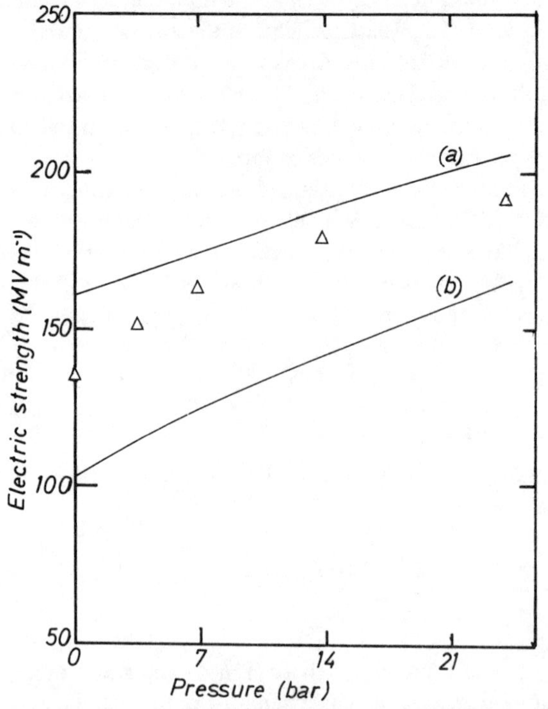


FIG. 3.22. Variation of breakdown strength of n-hexane with hydrostatic pressure. △, experiments of Kao and Higham (1961); — eqn. (3.16): (a) $R = 100$ Å; (b) $R = 250$ Å (after Krasucki 1966).

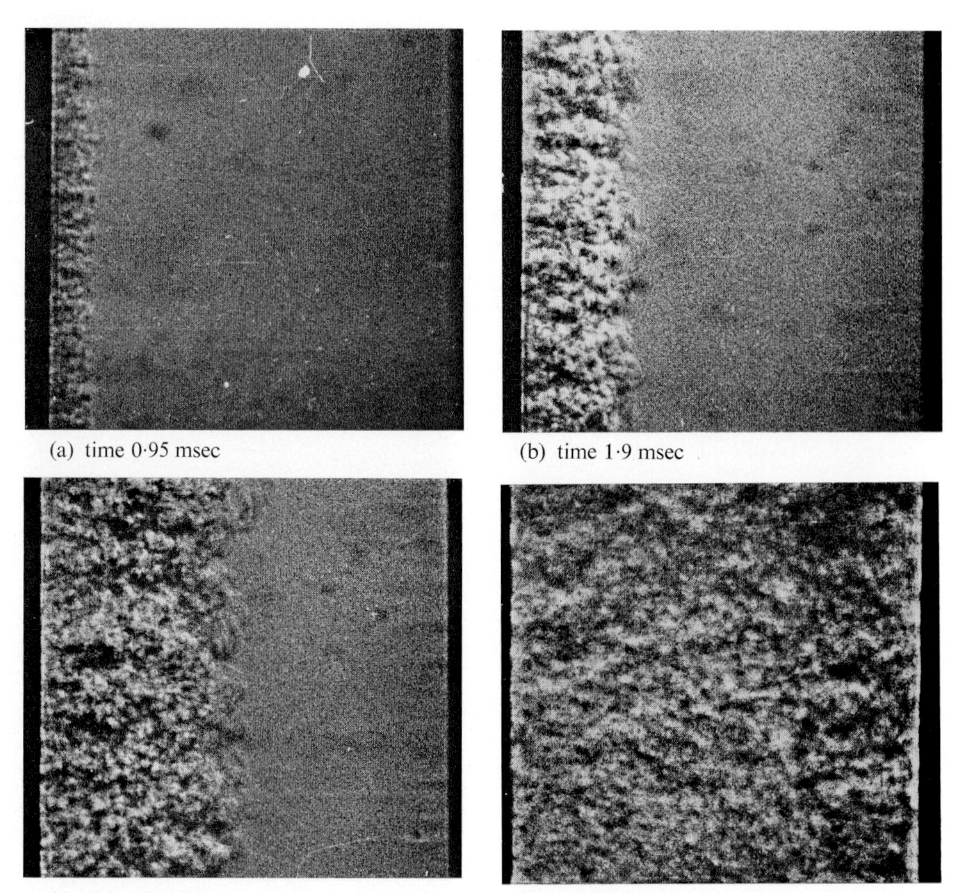

(a) time 0·95 msec

(b) time 1·9 msec

(c) time 2·9 msec

(d) steady state

PLATE 1: Schlieren photographs of the temporal propagation of turbulence in nitrobenzene after a step voltage of 18 kV is applied to electrodes 0·056 m apart. The injecting electrode (anode) is on the left side of each frame (after Hopfinger and Gosse 1971).

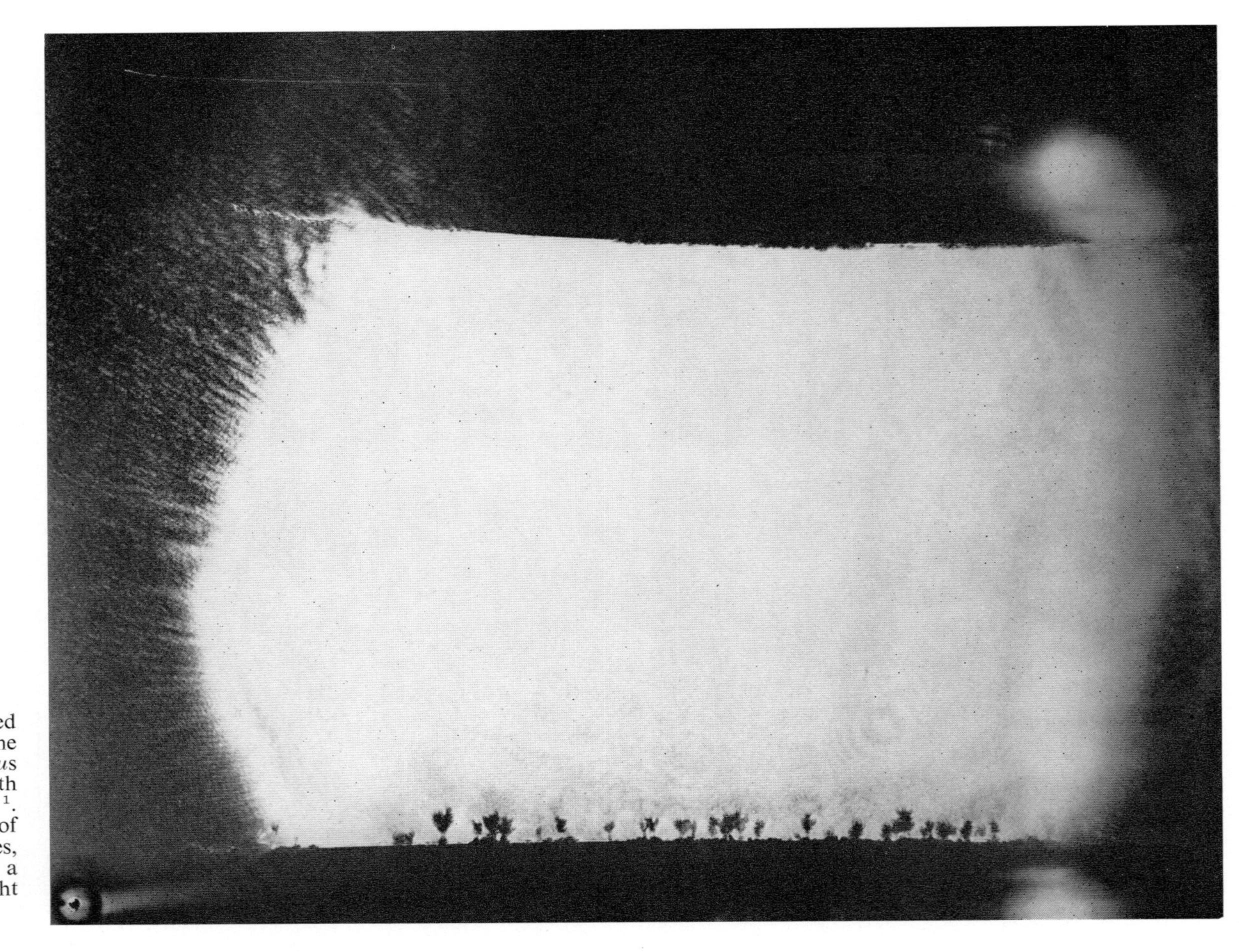

PLATE 2: Laser-illuminated Schlieren photograph of $n$-hexane between brass electrodes, 0·18 $\mu$s before breakdown. Gap length 4 mm. Applied field 62·5 $MVm^{-1}$. The picture shows a number of cavities, with irregular shapes, growing near the cathode, and a discharge commencing on the right (after Thomas 1973b).

particles may influence the electric strength but it is difficult to envisage that all breakdown measurements were made in liquids containing only particles with radii between 100 Å to 250 Å. Most investigators could not filter out particles less than $10^4$ Å in size (section 3.8) whilst many of the particles produced by spark erosion in argon formed clusters approximately $5 \times 10^3$ Å in size (Bucklow and Drain 1964). Inserting this value for $R$ in eqn (3.16) yields a strength of 34 MV $m^{-1}$ for n-hexane which is much lower than the values normally quoted in the literature. If we use Edward's maximum strength of 246 MV $m^{-1}$ for n-hexane (Table 3.8.2) then $R$ is approximately 40 Å. We are now approaching particle sizes not much greater than several molecular diameters. A detailed appraisal of Krasucki's theory has been given by Coelho and Gosse (1970). Mirza *et al.* (1970) have attempted to improve on the theory by considering the distortion of a bubble, formed after the zero pressure criterion is reached, but their calculated variation of strength is much too large and depends critically on bubble size.

The suggestion by Krasucki that cavities do initiate breakdown has received added support through the recent experimental work of Thomas (1973 *a*). Using a novel laser-illuminated Schlieren optical system a series of photographs was taken of small, local, density changes in n-hexane over a pre-determined interval during a breakdown event. The new feature discovered is that a number of opaque regions, approximately $10^{-4}$ m across, grow near the cathode about 0.2 $\mu$s before breakdown, as shown in Plate II (Thomas 1973 *b*). These regions, classed as cavities by Thomas, grow over a period of typically one more $\mu$s until one of them develops into a highly irregular shape before breakdown. Of greatest significance perhaps is the fact that in several hundred observations no cavities were seen which were much larger than those shown in Plate II. Thus, Thomas has concluded that, although a cavity may initiate a breakdown, in order to propagate it, other processes such as collision ionization at the cavity–liquid interface are required.

Following on Krasucki a zero pressure criterion for cavitation was developed by Thomas but it included several extra terms and was written as

$$P_c + P_{vp} = P_{es} + P_{st} + P_h. \qquad (3.17)$$

It is assumed that after electron injection into the liquid a so-called 'coulombic pressure' is produced which lifts the liquid off the cathode surface and is given by

$$P_c = \tfrac{1}{2}\,\epsilon_r\,\epsilon_0 \left\{ (mE_0)^2 - E_c{}^2 \right\}, \qquad (3.18)$$

where $m$ is a magnification of the applied field $E_0$ near the cathode, $E_c$ is the cathode field and $\epsilon_r$ is the relative permittivity of the liquid. The vapour pressure in the cavity is expressed as:

$$P_{vp} = K_1 \exp \left[ \frac{-K_2}{T + \Delta T} \right], \tag{3.19}$$

where $K_1$ and $K_2$ are constants for a chosen liquid and $\Delta T$ represents the rise in its temperature $T$. An expression for an electrostrictive pressure was developed which, it is claimed, in contrast to Krasucki, restricts lift-off of the liquid from the cathode, and is given by

$$P_{es} = \tfrac{1}{3}\, \epsilon_0\, (\epsilon_r - 1)(2\epsilon_r + 1) E_c^{\,2}. \tag{3.20}$$

The other terms on the right-hand side of eqn (3.17) refer to the surface tension, and hydrostatic, pressures. Breakdown is assumed to occur when cavitation takes place and therefore using eqns (3.17) to (3.20) we obtain

$$\tfrac{1}{2}\epsilon_r\, \epsilon_0\, (mE_s)^2 = \tfrac{1}{6}\, \epsilon_0\, (4\epsilon_r^{\,2} + \epsilon_r - 2) E_c^2 + \frac{2\sigma}{R} + P_h - K_1 \exp\left(\frac{-K_2}{T + \Delta T}\right), \tag{3.21}$$

where $E_s$ is the value of $E_0$ at breakdown. According to Thomas the major improvement in eqn (3.21) over eqn (3.16) is the inclusion of the cathode field $E_c$, as it represents a significantly large field. Particle effects are ignored completely but $R$ in eqn (3.21) is ascribed to the radius of a cathode asperity and it is shown that the surface tension pressure is negligibly small.

In order to find a solution to eqn (3.21) a Schottky-type emission equation was used to relate $E_c$ to the applied field $E_0$ and $\Delta\phi$, the work function for a cathode–liquid interface. Eqn (3.21) was then solved using iterative techniques. The procedure was to use known data for the physical constants of a liquid and to vary $\Delta\phi$ until agreement was obtained between the theoretical and experimental values of $E_s$. Other experimental parameters could then be varied to see if eqn (3.21) predicted the correct dependence. The technique was applied to n-hexane and Fig. 3.23 shows that for $e\Delta\phi = 0.9$ eV the predicted variation of strength with pressure is in excellent agreement with experimental results. The variation of strength with temperature and spacing was also found to be close to the theoretical curves. The value for $e\Delta\phi$ is close to the 0.94 eV computed by Morant (1960) for the total potential barrier to electron emission between a metal and a liquid, but is much lower than the 2 eV to 3 eV measured photoelectrically by Holroyd and Allen (1971). Using these values in eqn (3.21) would seriously alter the calculated values for $E_s$, as it is very sensitive even to small changes in $\Delta\phi$. Nevertheless, the experimental work of Thomas represents a step forward in our knowledge of the final stages of breakdown, but the factors responsible for the initiation and growth of a cavity still remain to be identified. The theoretical work contains some improvement over earlier models but the concepts of 'coulombic' and 'electrostrictive' pressures as used by Thomas are open to question.

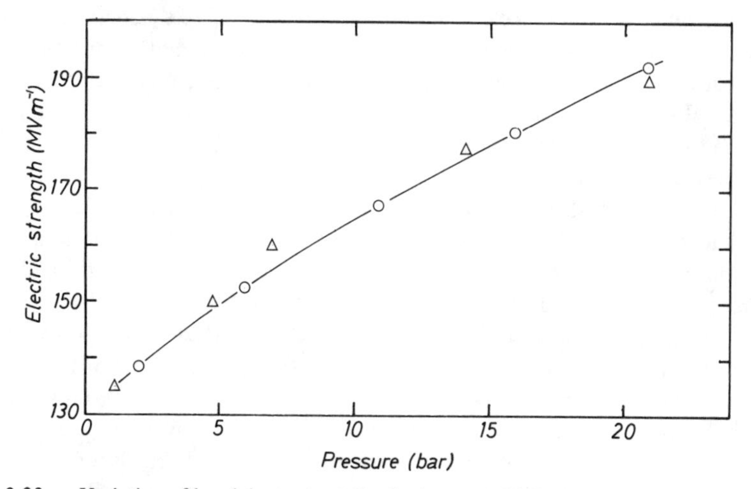


FIG. 3.23. Variation of breakdown strength of n-hexane with hydrostatic pressure. △, experiments of Kao and Higham (1961); -o-o, eqn (3.22) with $e\Delta\phi =$ 0.9 eV and $m = 4$ (after Thomas 1973*a*).

To conclude this sub-section concerned with theoretical explanations of pressure effects, it is appropriate that we comment briefly on the effects of an electric field on the shape of a bubble. In a theoretical analysis of electrically-induced stresses in dielectric liquids Scaife (1973) has conclusively shown that a spherical bubble in a non-conducting fluid becomes more oblate when subjected to an axial electric field. In marked contrast to this statement, several investigators have presented experimental evidence showing the elongation of a bubble in the direction of the field. The discrepancy between theory and experiment appears to arise from the fact that, under experimental conditions, the host liquid is in a conducting state, and a bubble may adopt a more spheroidal shape under the influence of surface charges at the bubble–liquid interface. For example, Garton and Krasucki (1964) and Kao (1965) have observed the elongation of an air bubble in a low viscosity oil, but the lowest applied stresses were 1 MV $m^{-1}$ and 2.5 MV $m^{-1}$, respectively. At these stresses the liquid was most probably in a conducting state. It would appear, therefore, that any theory of breakdown in liquids which is based on the growth and deformation of bubbles should allow for the effects of charge accumulated on the bubble surface. To date, such a theory is not available.

## 3.10. Pulse measurements

Throughout this chapter we have illustrated the effects of various experimental parameters on breakdown by featuring results which were obtained under conditions of direct or pulse voltages. Obviously, the use of pulse stresses of short duration to measure electric strength introduces an additional time factor which we now want to consider. Pulse breakdown measurements were undertaken in an effort to reduce thermal and space charge effects associated with direct voltage tests, and also to obtain information on ion motion and time to breakdown in liquids. However, most investigators did not appreciate the statistical implications of the pulse technique. The main object in this section will be to discuss pulse measurements on the basis of a statistical theory of breakdown.

As early as 1934 Strigel used statistical arguments to analyse the time-to-breakdown in transformer oil, whilst extreme-value statistics have been used to explain the breakdown of paper capacitors (Epstein and Brooks 1948) and the dependence on electrode area of the strength of transformer oil under 60 Hz alternating stress (Weber and Endicott 1956). The statistical nature of the time to breakdown using short pulses has been reported by Hancox (1956) for transformer oil, and by Saxe and Lewis (1955), using step-function pulses, for n-hexane. Several years elapsed before Ward and Lewis (1960) gave a preliminary account of a statistical treatment of pulse measurements. In the last decade or so many conflicting results in the literature have been successfully explained by a statistical model, and we shall see that there is striking agreement between theory and experiment.

### *3.10.1. The statistical theory*

When a dielectric liquid is suddenly subjected to a high electric stress, it retains its insulating properties for a time before the formation of a low impedance path between the electrodes. The statistical model divides this time into two components, a statistical lag $t_s$ and a formative lag $t_f$. The time $t_s$ is defined as the period between the application of stress and the appearance of an initiating event which will eventually lead to breakdown; $t_f$ is the time required to complete a breakdown after the appearance of the critical initiating event. These definitions are quite general, and no physical processes need be postulated for them. The only assumption required is that initiating events occur randomly at a mean rate $f(E)$ when a stress $E$ is applied to the electrode–liquid system.

Because it is easier to interpret the effects of wave-shape, pulse measurements have generally been made with step-function pulses or rectangular pulses of short duration, rather than the '1/50' type of pulse which is usually used for industrial tests on insulating liquids. Statistical evidence is instantly observed with the first type of pulse, as time lags to breakdown can be measured directly for a fixed value of stress. With the second type, the usual testing procedure, generally called a multiple-pulse method, is to apply pulses of fixed duration at increasing values of stress until breakdown occurs. This yields an electric strength for a particular pulse duration and although there is a spread in results the method tends to obscure the fact that a statistical lag is present.

If a liquid is subjected to $N_0$ applications of a step-function stress $E$, and $N_t$ breakdowns occur at a time greater than $t$, then it is easily shown (Lewis and Ward 1962) that

$$\ln (N_t/N_0) = -f(E)(t - t_{\mathrm{f}}). \tag{3.22}$$

If $f(E)$ is constant during the sequence of $N_0$ measurements then $\ln (N_t/N_0)$ is a linear function of time and eqn (3.22) is usually represented as a Laue plot. The mean statistical time lag is given by

$$\bar{t}_{\mathrm{s}} = \frac{1}{f(E)}, \tag{3.23}$$

and $t_{\mathrm{f}}$ is given by the intercept on the time axis at which $\ln (N_t/N_0) = 0$. This analysis, first suggested by von Laue (1925), has often been used in time lag measurements in gases (Meek and Craggs 1953). Accurate determinations of $f(E)$ from eqn (3.22) require that $N_0$ should be large and $f(E)$ constant during a sequence but it is difficult to satisfy both these conditions because of the damage caused by repeated discharges. When short pulses of duration $T$, at a stress $E$, are used the probability $p(E)$ of breakdown during $T$ is, from eqn (3.22)

$$p(E) = 1 - \exp\left\{-f(E)(T - t_{\mathrm{f}})\right\}. \tag{3.24}$$

Rearranging eqn (3.24) gives,

$$f(E) = \frac{(T - t_{\mathrm{f}})^{-1}}{\ln [1 - p(E)]} \tag{3.25}$$

Therefore, by calculating $p(E)$ from a large number of breakdowns at various levels of stress and different pulse durations, the experimental variation of $f(E)$ against stress can be derived. Also, eqn (3.24) can be developed to take account of the testing procedures used by different investigators. The mathematical analysis pertinent to this part of the theory can be found in the articles by Lewis and Ward (1962) and by Brignell (1966). Only certain statistical implications arising from the use of the multiple-pulse technique will be mentioned here. Thus, in a particular series of measurements with this test method, numerous factors will affect the measured electric strength. These can include the starting level $E_0$ in the series, the number $\lambda$ of pulse applications at each level of stress, the increments of stress $\Delta E$, pulse duration, electrode material and spacing, and the liquid itself. For instance, an increase of $E_0$, $\lambda$, or $\Delta E$ will cause a decrease in average electric strength and in the scatter of measurements, since the probability of breakdown is now greater for the higher stress and larger number of pulse applications. These facts have been clearly demonstrated both experimentally (Ward and Lewis 1960) and theoretically (Metzmacher and Brignell 1968 *b*).

### *3.10.2. Effect of pulse duration*

Crowe (1956) has measured the average electric strength for a range of hydrocarbon liquids at pulse durations up to 3 μs, as shown in Fig. 3.24. The 'knee-shaped' curves are typical of results obtained with short duration pulses. Beyond a certain time $T_0$ the strength $E_s$ (Fig. 3.24) is virtually time-independent but it

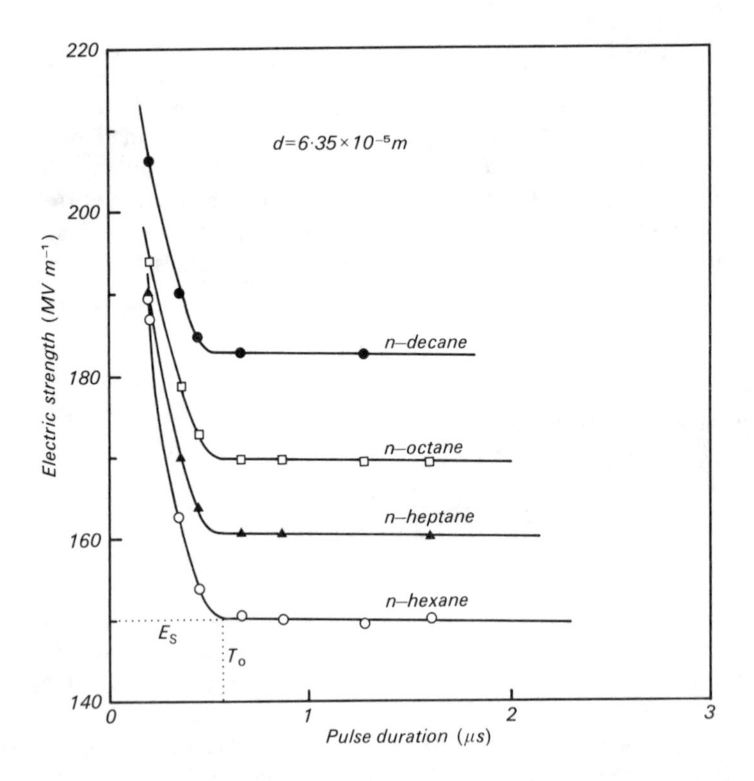


FIG. 3.24. The dependence of the electric strengths of a series of straight-chain hydrocarbon liquids upon pulse duration for a gap length of 63.5 μm (after Crowe 1956).

increases rapidly for $T<T_0$ and is practically insensitive to liquid structure in this range. The observations by Crowe confirmed the findings of Edwards (1951), whilst similar results have since been reported by Kao and Higham (1961), although there were small variations between the values of $T_0$ measured by all these workers. In direct contrast, Goodwin and MacFadyen (1953) have shown a definite correlation between molecular structure and $T_0$, which varied from approximately 1 $\mu$s for n-hexane to nearly 12 $\mu$s for n-tetradecane. An explanation of this large discrepancy has never been found but Goodwin and MacFadyen have stated that 'the breakdown time $T_0$ was varied for most of the liquids by using several different gap lengths, and in the case of benzene, different temperatures'. Since $T_0$ does increase with spacing their results may only be a representation of this fact. Despite the disagreement in the value of $T_0$ all these workers assumed that for $T>T_0$ and $E=E_s$ (Fig. 3.24) breakdown invariably occurred at $T_0$. Consequently, for $T<T_0$ the entire pulse duration was seen as a measure of the formative time to breakdown and the statistical lag was assumed to be zero. However, Lewis and Ward (1962) have shown that a significant statistical time lag exists in the breakdown process, which led them to propose an interpretation of pulse measurements which is very different from that adopted by earlier workers. Having determined the stress-dependence of the function $f(E)$ experimentally for a sphere–sphere system in n-hexane (eqn 3.25), and with $t_f = 0.1$ $\mu$s, Lewis and Ward (1962) used a statistical analysis to predict the shape of the curves in Fig. 3.24. The important result was that $T_0$, which was $\sim 4$ $\mu$s, was a consequence of the inherent statistical nature of pulse testing and was not related to $t_f$. Also, it was shown that the portion of the curves of $E$ versus $T$ at pulses $T<T_0$ was a result of the statistical availability of electrons from the cathode, whilst for $T>T_0$ the strength was chiefly determined by liquid properties. Thus, the time $T_0$ is a useful indication of cathode activity, but the point of transition from liquid to cathode control is not expected to be well defined. A consistent explanation of many results using pulse techniques can be given by the statistical argument (Lewis and Ward 1962). For instance, a correlation between $T_0$ and cathode work function, observed by Goodwin and MacFadyen (1953), is easily explained, whilst the results in Fig. 3.24 tend to confirm that the strength is dependent on liquid properties for pulses $T>T_0$.

Of the liquefied gases, breakdown measurements using pulse voltages only exist for liquid argon (Gallagher and Lewis 1964 *b*). It was found that the general behaviour of argon under pulse conditions was similar to that of liquid hydrocarbons except that $T_0$ was much larger and was $\sim 10$ $\mu$s for a spacing of 50 $\mu$m. On the basis of a statistical model, it was deduced that cathode emission was much less in liquid argon than in similar experiments on hydrocarbons. The low level of natural conductivity under direct stress, and the absence of any prestressing effects (section 3.7) on the pulse strength in argon is in agreement with this deduction. The variation of the breakdown voltage with pulse duration for a point–sphere system showed that with a point cathode $T_0$ was $\leqslant 2\ \mu s$ at a spacing of 50 $\mu m$. Statistically, this is to be expected because, for the point cathode, $f(E)$ is likely to be large, and the mean statistical time lag will be small. With the sphere as cathode, the breakdown voltages were greater, and the dependence on pulse duration tended to follow the pattern for a sphere–sphere geometry.

The association of a statistical lag with cathode emission, and the increase in $T_0$ with electrode spacing, were seen as major objections to a statistical theory of breakdown (Sharbaugh and Watson 1962). However, using a statistical model, with the aid of a few extensions to the original theory, Metzmacher and Brignell (1968 *b*) have predicted correctly the dependence of strength on spacing, pulse wave-shape, and test procedure. Moreover, in an experimental investigation of time lags, Brignell and Metzmacher (1971) have shown that the mean statistical time lag, or $f(E)^{-1}$, was independent of spacing, dependent on electrode material, and directly proportional to the area of the plane electrodes (Fig. 3.25). These facts suggest most strongly that breakdown initiating events are located on electrode surfaces.

### *3.10.3. Statistical and formative lags*

Fig. 3.26 shows typical sequences of time lag measurements in liquid argon using step-function pulses at stresses of 140 and 170 MV $m^{-1}$. At both stresses there was rapid conditioning so that the time lags became longer until, at the end of each sequence, no breakdown occurred at all, indicating time lags at least as long as 150 $\mu s$, the equivalent 'rectangular' pulse duration of the step-function voltage. It is interesting to note that time lags appear to follow a pattern of conditioning similar to that in breakdown measurements. These trends have also been noticed for n-hexane (Lewis and Ward 1962). The disagreement between the time lags obtained with 'short-pulse' and 'step-function pulse' test methods has been ascribed to the use, in the former case, of fresh electrode surfaces for each breakdown (Gallagher and Lewis 1964 *b*). A freshly-prepared cathode is likely to exhibit a high activity (section 3.4), so that $f(E)$ would be large, and time lags would be small, as indicated by the first measurement in Fig. 3.26. The short time lags measured by Kao and Higham (1961), using a single-shot technique and an unblemished electrode surface, are consistent with the above picture and are not contrary to the findings of Ward and Lewis (1960), as was suggested by Sharbaugh and Watson (1962).

Laue plots from sequences of time lag measurements have yielded values for $f(E)$ of $0.5 \times 10^6$ events $s^{-1}$ at 180 MV $m^{-1}$ in argon, and $0.77 \times 10^6$ events $s^{-1}$ at only 120 MV $m^{-1}$ in n-hexane. The smaller value in argon suggests that, with short pulses, $T_0$ should be greater than in n-hexane, in agreement with experiment. However, $f(E)$ is not only a stress dependent constant, as was first assumed by Lewis and Ward (1962), but it is also time dependent, as was shown by Brignell (1966), and by Gzowski (1968). Thus $f(E)$ can vary from $10^5$ events $s^{-1}$ in the

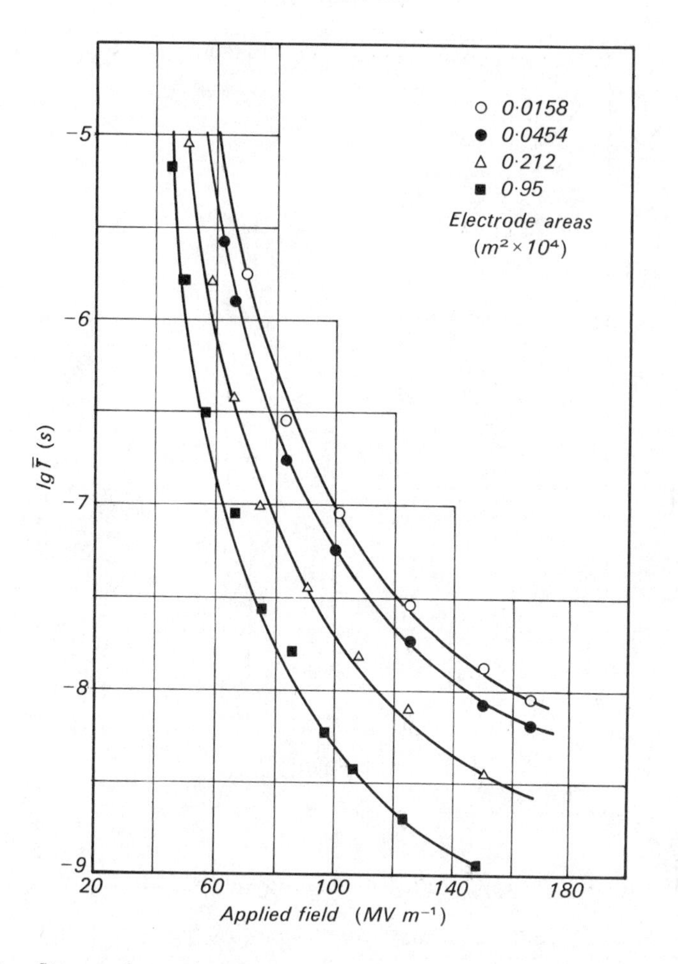


FIG. 3.25. Stress, and area, dependence of the mean breakdown time lag in n-hexane for brass electrodes at a gap length of 120 μm (after Brignell and Metzmacher 1971).

'microsecond' domain to 0.2 events $s^{-1}$ in the 'second' domain. Nevertheless, a statistical model is still valid, and it has been used successfully to analyse time lag measurements which have now been extended over nine decades of time; from the 'second' region (Brignell 1966) to the 'nanosecond' region (Beddow and Brignell 1965, 1966).

The formative lag $t_f$ was considered a constant in the early presentation of the statistical model, and Laue plots for hexane and argon indicated that it was $\leqslant 100$ ns. However, more recent results of breakdown measurements in the nanosecond region have shown that $t_f$ is, in fact, a distributed quantity. Since these experiments require high electrical stresses, $t_s$ is small and $t_f$ is the

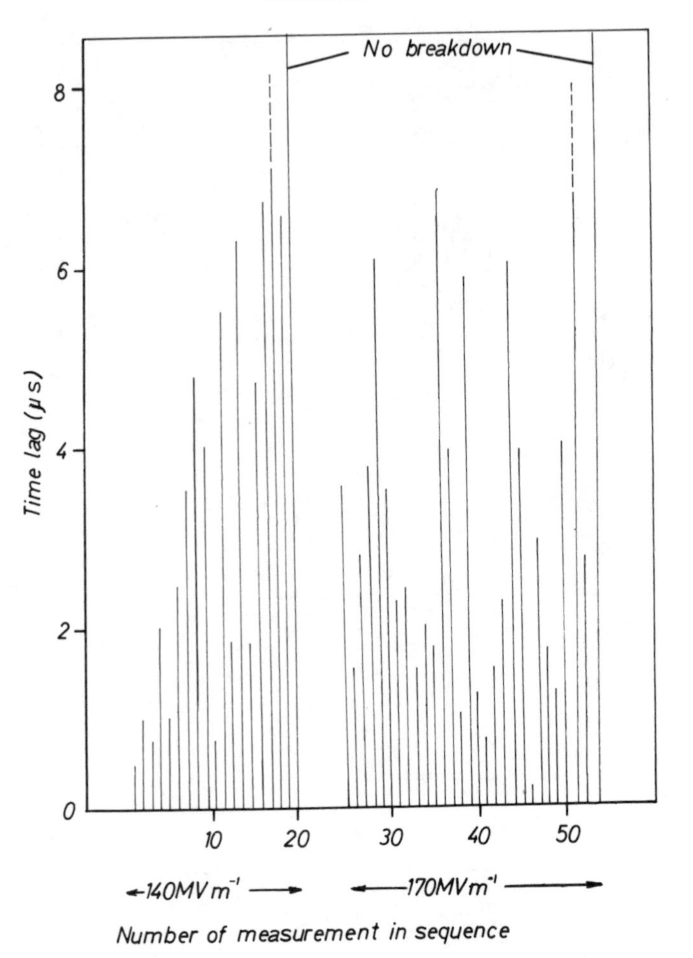


FIG. 3.26. Sequence of time lags at two stresses in liquid argon. Gap length 50 μm. Dotted lines indicate time lags beyond linear scale reading of recording instrument (after Gallagher and Lewis 1964*b*).

dominant component of the time lag. Beddow and Brignell (1965, 1966) were the first workers to examine this aspect of liquid breakdown. They have shown that at high stresses $t_f$ can be described by a Gaussian distribution function which may be written as

$$P(t_f) = \left(\frac{1}{\sigma\sqrt{2\pi}}\right) \exp\left[-(t-t_0)^2 / 2\sigma^2\right], \tag{3.26}$$

where $t_0$ is the mean of $t_f$, $\sigma$ is the standard deviation, and $P(t_f)$ is the probability density function. Mean formative times of 20.2 and 5.3 ns were obtained at stresses of 330 and 480 MV $m^{-1}$ for spacings of 31 and 36 $\mu$m, respectively, with individual values of $t_f$ sometimes lasting only 1 ns. The exponential and normal distributions for $t_s$ and $t_f$ have been confirmed by Palmer and House (1972 *a*). These authors also recorded the light output during the formative process, and they have shown that the time lag to the first light pulse yields a straight line Laue plot, in agreement with the concept of an exponentially distributed statistical lag. Felsenthal (1966) has measured normally distributed time lags at stresses between 200 and 400 MV $m^{-1}$ and at a spacing of 61 $\mu$m in n-hexane, transformer oil, paraffin oil, distilled water, and various synthetic liquids. On the other hand, Rudenko and Tsvetkov (1966) have obtained a linear Laue plot for formative lags but they neglected lags less than 2 ns; Beddow and Brignell (1966) have suggested that a droop in the applied voltage wave form and the method of plotting results may be other sources of error. Ward (1967) has reviewed all these measurements. Beddow, Brignell, and House (1968) have illustrated the probable stress dependence of the average time lag, and its distribution function, as shown in Fig. 3.27. In region 1a at low

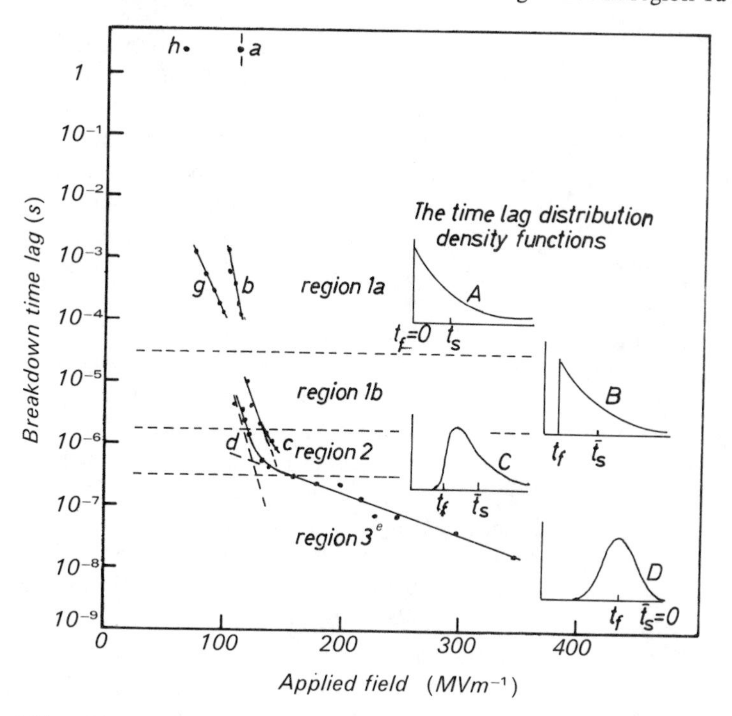


FIG. 3.27. The stress dependence of the time lag and its distribution function. Experimental points: *a, h* Brignell (1964); *b, c, g.* Lewis and Ward (1962); *d, e* Beddow (1968) (after Beddow, Brignell, and House 1968).

stresses, $t_f$ is negligible, and the time lag has an exponential distribution, which is retained in region 1b at slightly higher stresses, where $t_f$ appears as a small shift on the time axis. In region 3 at very high stresses, the time lag is normally distributed and entirely formative in nature. In the transition region 2 at intermediate stresses, $t_s$ and $t_f$ are of comparable magnitude, and the density function is got by direct convolution of the exponential and normal distributions.

Recently, Bell, Rogers, and Guenther (1972), apparently unaware of the work of Brignell and his colleagues, have investigated nanosecond breakdown in n-hexane for spacings between 700 to 4100 $\mu$m and stresses of 80 to 160 MV m$^{-1}$ These applied stresses span all of regions 1 and 2, but little of 3, in Fig. 3.27, and, consequently, the average time lag was likely to be mostly statistical rather than wholly formative, as was assumed. Shorter time lags were measured at much larger spacings and lower stresses than used by Felsenthal (1966), and this apparent contradiction was attributed to the fact that their electrode area was $\sim 10^3$ greater. On statistical arguments, it was unrealistic for Bell *et al.* (1972) to relate time lags, which they considered as formative, to electrode area, but the results, though misinterpreted, support the time lag measurements already shown in Fig. 3.25.

We must conclude from our discussions throughout this section that a statistical analysis provides the proper means to assess results obtained with pulse voltages. It is ironic that a relatively simple mathematical technique can be used to predict accurately the influence on electric strength of complex factors such as pulse duration and shape, test procedure, electrode spacing etc., whilst shedding little light on the physical mechanisms of breakdown. Nevertheless, the measurements serve to highlight two important points. Firstly, the breakdown process in liquids can operate in times of a few nanoseconds or less, and secondly, it is initiated at the electrodes.

It has been suggested (Lewis and Ward 1962; Gallagher and Lewis 1964 *b*) that breakdown is initiated by an extra large, localized, burst of electrons from certain active sites on the cathode where emission momentarily becomes easy. $f(E)$ would be an indication of the number of such sites on the surface. The migration of positive ions or charged dust particles to the cathode would promote these bursts. Krasucki (1966, 1971) has related $t_s$ to the transit time of a particle between the electrodes and appears to imply that $T_0$ in Fig. 3.24 is a measure of the transit time. However, Coelho and Gosse (1970) have pointed out that this model would yield a distribution function for $t_s$ having a maximum at the average transit time, whereas the observed functions are decreasing exponentials (Fig. 3.27). Also, unpublished calculations in the author's laboratory of particle transit times in argon and n-hexane have shown them to be almost identical. Therefore, based on Krasucki's model, $T_0$ should be the same for these liquids. This is contrary to the experimental results.

With regard to $t_f$, it is difficult to imagine bubble mechanisms operating in times of a few ns, especially as estimates of 100 ns have been made for the growth time of a vapour bubble in n-hexane (Krasucki 1966). It is also significant that the effect of pressure on electric strength disappeared for times less than about 300 ns (Fig. 3.20). To explain a breakdown process which takes place within a few ns, Felsenthal (1966) postulated that $t_s$ was zero, and that a discharge occurred when a single electron avalanche traversed the gap. This model yielded a charge carrier mobility of $8.7 \times 10^{-6}$ m$^2$ V$^{-1}$ s$^{-1}$ at a stress of 200 MV m$^{-1}$, which is in excellent agreement with the value for $\mu_e$ of $9 \times 10^{-6}$ m$^2$ V$^{-1}$ s$^{-1}$ measured directly by Schmidt and Allen (1970 *a*), at a stress of 8.3 MV m$^{-1}$. However, Palmer and House (1972 *a*), in an effort to separate an actual initiating event from post-breakdown arc formation, have used a streak-camera, operating at speeds many times faster than previous investigations, to show that the rudimentary breakdown bridges the gap with a velocity of, at least, $6 \times 10^6$ m s$^{-1}$. As we have seen in sub-section 1.4.1, the drift velocity of electrons in many hydrocarbon liquids is a linear function of stress up to 14 MV m$^{-1}$, the highest field used (Schmidt and Allen 1970 *a*). Now, since Palmer and House (1972 *a*) used a stress of 61.5 MV m$^{-1}$, a linear extrapolation of the drift velocity data would yield an electron velocity in their experiments of $5.54 \times 10^3$ m s$^{-1}$. This value is still three orders of magnitude lower than their estimate of the discharge velocity. It is difficult to conceive of any physical process in liquids which would allow electrons to increase their drift velocities by a factor of $10^3$ for a stress increment of only some fifty MV m$^{-1}$. Therefore, the concept of breakdown via a single avalanche must be discarded. It would appear that, at last, in the field of liquid dielectrics, we can use the results of mobility and pre-breakdown measurements to elucidate, in a quantitative way, some phenomena of breakdown.

Based on their measurements of nanosecond time lags, Bell *et al.* (1972) have proposed a mechanism of streamer formation, of which much is known in studies of gas and lightning discharges, and which can propagate in liquids at velocities of $\sim 10^6$ ms$^{-1}$. According to these authors, the critical avalanche needed to sustain streamer growth must start through electron emission from the cathode, or from the liquid near the anode. In either case, these electrons are confined to a small portion of the gap, and streamer formation and propagation take place at the head of the avalanche. A conducting channel between the electrodes is rapidly formed in this way. However, we must remember that streamer mechanisms have been formulated primarily for gases, and it is unlikely that they can be applied unmodified to describe completely breakdown in liquids. Nevertheless, there are numerous references in the literature showing light generation and streamer-like processes originating in the vicinity of one or both electrodes (Washburn 1933; Liao and Anderson 1953; Komelkov 1962; Stekol'nikov and Ushakov 1966, Allan and Hizal 1974).

## 3.11. Summary

There is no single theory that explains all the results of breakdown experiments. Many of the theories that rely on physical models of breakdown can account for the correct trends in measurements. However, agreement between theoretical estimates and experimental values for the electric strength is obtained by recourse to adjustable parameters. In marked contrast, the statistical theory of breakdown under pulse voltages, which is based on a simple mathematical model, can provide striking confirmation of many results. It is somewhat ironic that a mathematical analysis can predict results accurately in view of the very complicated nature of the breakdown process.

Having assessed all the results presented in this chapter an attempt is made in Table 3.11.1 to separate the stages in the temporal development of a conducting breakdown channel for a symmetrical electrode configuration. In Table 3.11.1 statistical lags are ignored and the time scales for the different stages are only approximate. Once stage 1 is complete the subsequent picture is expected to be the same under direct and pulse stresses. The initiating event may be caused by field enhancement at asperities, positive ions or charged dust particles arriving at the cathode. It will also be influenced by oxide, gas, and space charge, layers at the cathode. Exceptionally fast streak-photographs of a discharge propagation in n-hexane have revealed that stages 1 and 2 of a breakdown probably involve electron and photon mechanisms. Thus, by interacting with the emitted electrons, the liquid molecules can determine the degree of ionization and excitation near the cathode. As the excited molecules decay to their ground states, photon absorption in the bulk of the liquid will also influence the amount of light generated in stage 2.

Table 3.11.1.
*Successive stages in the breakdown of a dielectric liquid.*

| Stage | Time Lag | Process | Effect |
|---|---|---|---|
| 1 | – | Initiating event, followed by a burst of electron emission from the cathode. | Excitation and ionization collisions near the cathode. |
| 2 | $1 \rightarrow 20$ns | Streamer formation from cathode to anode, followed rapidly by return streamer to the cathode. | Light generation throughout the gap. Dense space charge in a filamentary channel through the liquid. |
| 3 | $20\text{ns} \rightarrow 1\mu\text{s}$ | Energy fed into the liquid from the cathode. Energy extracted at the anode. | Formation of a low-density disturbance at the cathode. Growth of this disturbance. |
| 4 | $1\mu\text{s} \rightarrow \infty$ | Continuation of process 3. | Breakdown and arc formation. |

Hydrostatic pressure may affect stage 1 by altering any gas in equilibrium at the electrodes, but is more likely to retard, or even collapse, stage 3. This would require a higher stress to sustain growth of the disturbance, and it would reflect a pressure dependence on the strength. For non-uniform electrode geometries, stage 1 is probably determined by field emission, or ionization, at the point electrode, depending on its polarity. Subsequently, a full discharge will develop from the point by way of the processes in stages 2 to 4. A liquid can influence all four stages in Table 3.11.1. It is not surprising, therefore, that *electronic* and *bubble* theories of breakdown have yielded some agreement with experiment.

It is clear that much more work is required before a full understanding of the fundamental processes of breakdown can be achieved. It is doubtful, however, if more significant information about these processes can be gained from conventional discharge measurements. More likely, the mechanisms will be elucidated by examining events within a few ns before breakdown. Schlieren and streak-photography techniques have already helped to clarify some of these events. Advancement in these techniques will provide further valuable information about the breakdown mechanism. Future work should also include electron mobility measurements on ultra-pure liquids at fields approaching breakdown. Moreover, since the results of conduction and breakdown experiments refer to 'impure' test samples, many of the traditional experiments need to be repeated on ultra-pure hydrocarbons.

At present, we have a comprehensive knowledge of the basic mechanisms of conduction and breakdown in the gaseous and solid states. We are only beginning, however, to understand the corresponding processes in the liquid state. An exciting era of research into dielectric liquids lies ahead.

# International Conferences on Dielectric Liquids

1959 1st Conference at Philadelphia, U.S.A. Proceedings published as individual papers.

1963 2nd Conference at Durham University, England. Proceedings published as individual papers. For a report of the Conference see Morant (1963).

1968 3rd Conference at Grenoble, France. Proceedings published by C.N.R.S. Paris, 1970, ed. N.J. Felici, Publication No. 179.

1972 4th Conference at Dublin, Ireland. Proceedings published by Typografia Hiberniae, Dublin, 1972, ed. T.J. Gallagher. For a brief report of the Conference see Gallagher (1972).

1969 1st Conference on EHD phenomena in dielectric liquids. Proceedings published by MIT., Cambridge, Massachusetts.

1970 I.E.E. Conference on Dielectric materials, measurements and applications, at University of Lancaster, England. Proceedings published as I.E.E. Conference Publication No. 67.

*Note* To avoid repetition in the bibliography, papers presented at the above meetings are referred to by the location of the Conference only.

# Appendix

*Note 1* There is an extensive literature dealing with the theory of EHD in the proceedings of the conferences at M.I.T. (1969), Lancaster (1970), and Dublin (1972). The articles by Felici (1969, 1971 *a*) may also be consulted.

*Note 2* Much research is now devoted to the feasibility of using distilled $H_2O$ as a dielectric in megajoule energy-density devices (cf. the article by Abramyan, Kornilov, Lagunov, Ponomarenko, and Soloukhin (1972).

*Note 3* Good discussions of the Kerr effect can be found in the articles by Mueller (1941), and by Cassidy and Cones (1969).

*Note 4* A comprehensive treatment of Schlieren methods is given in the monograph by Holder and North (1963).

*Note 5* The Ramsauer effect in rare gases is fully discussed by Massey and Burhop (1969): experimental results on pp. 24–30, 47–50; theoretical discussion on pp. 401–7.

*Note 6* Boltzmann's classic book '*Vorlesungen über Gastheorie*' has been translated into English by Brush (1964).

*Note 7* The liquids studied by Freeman and his colleagues include: alkane isomers (Dodelet and Freeman 1972), methyl-substituted propanes (Fuochi and Freeman 1972), liquefied gases and fluoromethanes (Robinson and Freeman 1973 *a, b*), and olefins (Dodelet, Shinsaka, and Freeman 1973).

*Note 8* For a discussion of Raman spectra confer chapter 3 of Herzberg (1945).

*Note 9* An extensive account, by Davis, H.T. and Brown, R.G., of the experimental and theoretical work on low energy electrons in non-polar liquids will be published in *Advances in Chemical Physics*. The author is indebted to Professor H.T. Davis for a preprint of this article.

# Bibliography


Abramyan, E.A., Kornilov, V.A., Lagunov, V.M., Ponomarenko, A.G., and Soloukhin, R.I. (1972). *Sov. Phys., Tech. Phys.* **16**, 983.

Adamczewski, I. (1937 *a*). *Ann. Phys.* **8**, 309.

—— (1937 *b*). *Acta. Phys. polon.* **6**, 432.

—— (1957). *Zeszyty nauk. Politech. Gdanski.* **3**, 3.

—— (1965). *Br. J. appl. Phys.* **16**, 759.

—— (1969). *Ionization, conductivity and breakdown in dielectric liquids.* Taylor and Francis, London.

Allan, R.N. and Hizal, E.M. (1974). *Proc. Instn. elect. Engrs.* **121**, 227.

Angerer, L. (1963). *Nature, Lond.* **199**, 62.

—— (1965). *Proc. Instn. elect. Engrs.* **112**, 1025.

Aniansson, G. (1955). *Phys. Rev.* **98**, 300.

—— (1961). *Trans. R. Inst. Technol., Stockholm,* No. 178.

Aplin, K.W. and Secker, P.E. (1972). *Proceedings of the Dublin Conference*, p. 158.

Atkins, K.R. (1959). *Phys. Rev.* **116**, 1339.

Atten, P. and Gosse, J.P. (1968). *Proceedings of the Grenoble Conference*, p. 325.

—— Moreau, R. (1969 *a*). *C. r. hebd. Séanc. Acad. Sci. Paris A* **269**, 433.

—— —— (1969 *b*). *C. r. hebd. Séanc. Acad. Sci. Paris A* **269**, 469.

—— —— (1970). *C. r. hebd. Séanc. Acad. Sci. Paris A* **270**, 415.

—— —— (1972). *J. Méc.* **11**, 471.

Bakale, G. and Schmidt, W.F. (1973 *a*). *Z. Naturforsch.* **28a**, 511.

—— —— (1973 *b*). *Chem. Phys. Lett.* **22**, 164.

Basseches, H. and Barnes, M.W. (1958). *Ind. Engng Chem.* **50**, 959.

Beck, G. and Thomas, J.K. (1972). *J. chem. Phys.* **57**, 3649.

Beddow, A.J. and Brignell, J.E. (1965). *Electron. Lett.* **1**, 253.

—— —— (1966). *Electron. Lett.* **2**, 142.

—— —— House, H. (1968). *Proceedings of the Grenoble Conference*, p. 99.

Bell, J.A., Rogers, R.R., and Guenther, A.H. (1972). *IEEE. Trans. Elect. Insul.* (*Inst. elect. electron. Engrs.*) **EI-7**, 78.

Belmont, M.R. and Secker, P.E. (1971). *J. Phys. D, Appl. Phys.* **4**, 956.

—— —— (1972). *J. Phys. D, Appl. Phys.* **5**, 2212.

Birlasakeran, S. and Darveniza, M. (1972 *a*). *Proceedings of the Dublin Conference*, p. 120.

—— —— (1972 *b*). *Proceedings of the Dublin Conference*, p. 124.

Blaisse, B.S., Van Den Boogart, A., and Erne, F. (1958). *Bull. Inst. Intern. du Froid, Commission 1, Delft, Annexe 1958*, p. 333.

—— Goldschvartz, J.M., and Slagter, P.C. (1970). *Cryogenics* **10**, 163.

Blank, C. and Edwards, M.H. (1960). *Phys. Rev.* **119**, 50.

Bohon, R.L. and Claussen, W.F. (1951). *J. Am. chem. Soc.* **73**, 1571.

Boone, W. and Vermeer, J. (1972). *Proceedings of the Dublin Conference*, p. 214.

Brière, G. (1964). *Br. J. appl. Phys.* **15**, 413.

—— Felici, N.J. (1960). *C. r. hebd. Séanc. Acad. Sci. Paris* **251**, 1004.
—— —— (1964). *C. r. hebd. Séanc. Acad. Sci. Paris* **259**, 3237.
—— Gaspard, F. (1968). *Chem. Phys. Lett.* **1**, 706.
—— Gosse, J.P. (1968). *J. chim. Phys.* **65**, 1341.
—— Rose, B. (1967). *J. chim. Phys.* **64**, 1720.
Bright, A.W., Makin, B., and Pearmain, A.J. (1969). *Br. J. appl. Phys.* (*J. Phys. D*) **2**, 447
Brignell, J.E. (1963). *J. scient. Instrum.* **40**, 576.
—— (1966). *Proc. Inst. elect. Engrs.* **113**, 1683.
—— Buttle, A.J. (1971). *J. Phys. D, Appl. Phys.* **4**, 1560.
—— House, H. (1965). *Nature, Lond.* **206**, 1142.
—— Metzmacher, K.D. (1971). *J. Phys. D, Appl. Phys.* **4**, 253.
Bruschi, L. and Santini, M. (1970). *Rev. scient. Instrum.* **41**, 102.
—— Mazzi, G., and Santini, M. (1970). *Phys. Rev. Lett.* **25**, 330.
—— —— —— (1972). *Phys. Rev. Lett.* **28**, 1504.
—— Mazzoldi, P., and Santini, M. (1966). *Phys. Rev. Lett.* **17**, 292.
—— —— —— (1967). *Phys. Rev.* **167**, 203.
Brush, S.G. (1964). *Lectures on gas theory.* University of California Press.
Bucklow, I.A. and Drain, L.E. (1964). *J. scient. Instrum.* **41**, 614.
Burnier, P.H., Moreau, J.L., and Lehmann, J.P. (1970). *Adv. cryogen. Eng.* **15**, 76.
Byatt, S.W. and Secker, P.E. (1968). *Br. J. appl. Phys.* (*J. Phys. D*) **1**, 1011.
Careri, G., Cunsolo, S., and Mazzoldi, P. (1964). *Phys. Rev. A* **136**, 303.
Cassidy, E.C. and Cones, H.N. (1969). *J. Res. natn. Bur. Stand. U.S.A.* (*Eng. Instrum.*) C **73**, 5–13.
Chadband, W.G. and Calderwood, J.H. (1972). *Proc. Instn. elect. Engrs.* **119**, 1661.
—— Wright, G.T. (1965). *Br. J. appl. Phys.* **16**, 305.
—— Coelho, R., and Debeau, J. (1971). *J. Phys. D, Appl. Phys.* **4**, 539.
Chong, P. and Inuishi, Y. (1960). *Tech. Rep. Osaka Univ.* **10**, 545.
Coe, G., Hughes, J.F., and Secker, P.E. (1966). *Br. J. appl. Phys.* **17**, 885.
Coelho, R. and Gosse, J.P. (1970). *Annls Phys.* **5**, 255.
—— Sibillot, P. (1969). *Nature, Lond.* **221**, 757.
—— —— (1970). *Rev. gen. Eléct.* **79**, 29.
Cohen, M.H. and Lekner, J. (1967). *Phys. Rev.* **158**, 305.
Conrad, E.E. and Silverman, J. (1969). *J. chem. Phys.* **51**, 450.
Craggs, J.D., Thorburn, R., and Tozer, B.A. (1957). *Proc. R. Soc. A* **240**, 473.
Crawley, J. and Angerer, L. (1966). *Proc. Instn. elect. Engrs.* **113**, 1103.
Crowe, R.W. (1956). *J. appl. Phys.* **27**, 156.
—— Sharbaugh, A.H., and Bragg, J.K. (1954). *J. appl. Phys.* **25**, 1480.
Cunsolo, S. (1961). *Nuovo Cim.* **21**, 76.
Dakin, T.W. and Berg, D. (1959). *Nature, Lond.* **184**, 120.
—— Hughes, J. (1968). *Annual Report of the 1968 Conference on electrical insulation and dielectric phenomena*, p. 68.
Darveniza, M. (1969). *Elect. Eng. Trans. I.E. Aust.*, September issue, 284–9.
—— Tropper, H. (1961). *Proc. phys. Soc.* **78**, 854.
Davidson, N. and Larsh, A.E. (1948). *Phys. Rev.* **74**, 220.
—— —— (1950). *Phys. Rev.* **77**, 706.
Davis, H.T., Rice, S.A., and Meyer, L. (1962 *a*). *J. chem. Phys.* **37**, 2470.
—— —— —— (1962 *b*). *J. chem. Phys.* **37**, 947.
—— —— —— (1962 *c*). *J. chem. Phys.* **37**, 1521.

—— Schmidt, L.D., and Minday, R.M. (1971). *Phys. Rev. A* **3**, 1027.
—— —— —— (1972). *Chem. Phys. Lett.* **13**, 413.
De Groot, K., Gary, L.P., and Jarnagin, R.C. (1967). *J. chem. Phys.* **47**, 3084.
Denholm, A. (1958). *Can. J. Phys.* **36**, 476.
Devins, J.C. and Wei, J.G. (1972). *Proceedings of the Dublin Conference*, p. 13.
Dewar, M.J. (1946). *J. chem. Soc.* **1**, 406.
Dey, T.H. and Lewis, T.J. (1968). *J. Phys, D, Appl. Phys.* **1**, 1019.
Dionne, V.E., Young, R.A., and Tomizuka, C.T. (1972). *Phys. Rev. A* **5**, 1403.
Doake, S.M. and Gribbon, P.W.F. (1971). *J. Phys. A, Gen. Phys.* **4**, 952.
Dodelet, J.P. and Freeman, G.R. (1972). *Can. J. Chem.* **50**, 2667.
—— Shinsaka, K., and Freeman, G.R. (1973). *J. chem. Phys.* **59**, 1293.
Durand, P. and Fournié, R. (1970). *Proceedings of the Lancaster Conference*, p. 142.
Edwards, W.D. (1951). *Can. J. Phys.* **29**, 310.
—— (1952). *J. chem. Phys.* **20**, 753.
Epstein, B. and Brooks, H. (1948). *J. appl. Phys.* **19**, 544.
Essex, V. and Secker, P.E. (1968). *Br. J. appl. Phys.* (*J. Phys. D*) **1**, 63.
—— —— (1969). *Br. J. appl. Phys.* (*J. Phys. D*) **2**, 1107.
Fallou, B., Galand, J., Bobo, J., and Dubois, A. (1969). *Bull. Inst. Intern. du Froid, Commission I, London, Annexe 1969*, p. 377.
Farazmand, B. (1961). *Br. J. appl. Phys.* **12**, 251.
Felici, N.J. (1967). *Rev. gén. Eléct.* **76**, 786.
—— (1969). *Rev. gén. Eléct.* **78**, 717.
—— (1971 *a*). *Direct Current* **2**, 147.
—— (1971 *b*). *Direct Current* **2**, 90.
Felsenthal, P. (1966). *J. appl. Phys.* **37**, 3713.
—— Vonnegut, B. (1967). *Br. J. appl. Phys.* **18**, 1801.
Filippini, J.C., Lacroix, J.C., and Tobazeon, R. (1970). *Proceedings of the Lancaster Conference*, p. 121.
—— Gosse, J.P., Lacroix, J.C., and Tobazeon, R. (1969). *C. r. hebd. Séanc. Acad. Sci. Paris B* **269**, 736.
—— —— —— —— (1970). *Proceedings of the Lancaster Conference*, p. 110.
Fisher, B.P. and Taylor, R.J. (1971). *J. Phys. D, Appl. Phys.* **4**, 1958.
Forster, E.O. (1962). *J. chem. Phys.* **37**, 1021.
—— (1964 *a*). *J. chem. Phys.* **40**, 86.
—— (1964 *b*). *J. chem. Phys.* **40**, 91.
—— (1967). *IEEE Trans., Elect. Insul.* (*Inst. elect. electron. Engrs.*) **EI-2**, 10.
—— (1972 *a*). *Proceedings of the Dublin Conference*, p. 148.
—— (1972 *b*). *Annual Report of the 1972 Conference on electrical insulation and dielectric phenomena*, p. 30.
Fournié, R. (1970). *Bull. Elect. Fr. Sér. B*, 53–64.
Fowler, R.H. and Nordheim, L.W. (1928). *Proc. R. Soc. A* **119**, 173.
Freeman, G.R. (1973). *Can. J. Phys.* **51**, 1686.
Frommhold, L. (1968). *Phys. Rev.* **172**, 118.
Fueki, K., Feng, D.F., and Kevan, L. (1972). *Chem. Phys. Lett.* **13**, 616.
Fuochi, P.G. and Freeman, G.R. (1972). *J. chem. Phys.* **56**, 2333.
Gachechiladze, N.A., Mezhov-Deglin, L.P., and Shal'nikov, A.I. (1970). *Sov. Phys. JETP Lett.* **12**, 159.
Galand, J. (1968). *C. r. hebd. Séanc. Acad. Sci. Paris* **266**, 1302.
Gallagher, T.J. (1962). Ph.D. Thesis. University of London.

—— (1968). *Proceedings of the Grenoble Conference*, p. 113.
—— (1972). *Physics Bulletin,* **23**, 731.
—— (1973). *Annual Report of the 1973 Conference on electrical insulation and dielectric phenomena* p. 503.
—— Lewis, T.J. (1964 *a*). *Br. J. appl. Phys.* **15**, 491.
—— —— (1964 *b*). *Br. J. appl. Phys.* **15**, 929.
Garben, B. (1972). *Proceedings of the Dublin Conference,* p. 144.
—— (1974). *Phys. Lett. A* **47**, 153.
Garton, C.G. and Krasucki, Z. (1964). *Proc. R. Soc. A* **280**, 211.
Gauster, W.F. and Schwenterly, J.W. (1973). *Oak Ridge National Laboratory Report* TM-4187.
Gerhold, J. (1972). *Cryogenics,* **12**, 370.
Germer, L. (1959). *J. appl. Phys.* **30**, 46.
Goldschvartz, J.M. and Blaisse, B.S. (1966). *Br. J. appl. Phys.* **17**, 1083.
—— —— (1970). *Bull. Inst. Intern. du Froid, Commission 1, Tokyo, Annexe 1970*, p. 231.
—— van Steeg, C., Arts, A.F.M., and Blaisse, B.S. (1972). *Proceedings of the Dublin Conference*, p. 228.
Goodstein, D.L., Buontempo, U., and Cerdonio, M. (1968). *Phys. Rev.* **171**, 181.
Goodwin, D.W. and MacFadyen, K.A. (1953). *Proc. phys. Soc. B* **66**, 85.
Gosling, C.H. and Tropper, H. (1964). *Proceedings of the IEE Conference on dielectric and insulating materials,*
Goswami, B.M., Angerer, L., and Ward, B.W. (1972). *Proceedings of the Dublin Conference,* p. 218.
Gray, E. and Lewis, T.J. (1965). *Br. J. appl. Phys.* **16**, 1049.
—— —— (1969). *Br. J. appl. Phys.* (*J. Phys. D*) **2**, 93.
Green, W.B. (1955). *J. appl. Phys.* **26**, 1257.
Grunberg, L. (1958). *Br. J. appl. Phys.* **9**, 85.
—— Wright, K.H.R. (1953). *Nature, Lond.* **171**, 890.
—— —— (1955). *Proc. R. Soc. A* **232**, 423.
Guizonnier, R. (1961). *J. electrochem. Soc.* **108**, 519.
—— (1968). *Proceedings of the Grenoble Conference* p. 251.
Gzowski, O. (1962 *a*). *Nature, Lond.* **194**, 173.
—— (1962 *b*). *Z. phys. Chem.* **221**, 288.
—— (1966). *Nature, Lond.* **212**, 185.
—— (1968). *Proceedings of the Grenoble Conference,* p. 131.
—— Terlecki, J. (1959). *Acta Phys. polon.* **18**, 191.
—— Liwo, J., and Piltkowska, J. (1968). *Proceedings of the Grenoble Conference,* p. 141.
—— Wlodarski, R., Hesketh, T.R., and Lewis, T.J. (1966). *Br. J. appl. Phys.* **17**, 1483.
Hakim, S.S. and Higham, J.B. (1961). *Nature, Lond.* **189**, 996.
Halpern, B. and Gomer, R. (1965). *J. chem. Phys.* **43**, 1069.
—— —— (1969 *a*). *J. chem. Phys.* **51**, 1031.
—— —— (1969 *b*). *J. chem. Phys.* **51**, 1048.
—— Lekner, J., Rice, S.A., and Gomer, R. (1967). *Phys. Rev.* **156**, 351.
Hancox, R. (1956). *Nature, Lond.* **178**, 1305.
—— (1957). *Br. J. appl. Phys.* **8**, 476.
—— Tropper, H. (1957). *Proc. Instn. elect. Engrs. A* **105**, 250.
Henshaw, D.G., Hurst, D.G., and Pope, N.K. (1953). *Phys. Rev.* **92**, 1229.
Henson, B.L. (1964). *Phys. Rev. A* **135**, 1002.

—— (1970). *Phys. Rev. Lett.* **24**, 1327.
Herzberg, G. (1945). *Molecular spectra and molecular structure*, Vol. II. *Infrared and Raman spectra of polyatomic molecules*, pp. 192–5. Van Nostrand, New York.
Hesketh, T.R. (1966). Ph.D. Thesis, University of London.
—— Lewis, T.J. (1969). *Br. J. appl. Phys.* (*J. Phys. D*) **2**, 557.
Hewish, T.R. and Brignell, J.E. (1972). *J. Phys. D, Appl. Phys.* **5**, 747.
Holbeche, T.H. (1956). Ph.D. Thesis, University of Birmingham.
Holder, D.W. and North, R.J. (1963). *Schlieren methods.* Notes on applied Science, No. 31. HMSO.
Holroyd, R.A. (1972). *J. chem. Phys.* **57**, 3007.
—— Allen, M. (1971). *J. chem. Phys.* **54**, 5014.
—— Tauchert, W. (1974). *J. chem. Phys.* **60**, 3715.
—— Dietrich, B.K., and Schwarz, H.A. (1972). *J. phys. Chem.* **76**, 3794.
Hopfinger, E.J. and Gosse, J.P. (1971). *Phys. Fluids* **14**, 1671.
House, H. (1957). *Proc. phys. Soc. B* **70**, 913.
Huang, K. and Olinto, A.C. (1965). *Phys. Rev. A* **139**, 1441.
Hughes, J.F. (1970). *J. chem. Phys.* **53**, 2598.
Hummel, A. and Schmidt, W.F. (1971). Report of HAHN-MEITNER Institute, HMI-B117, Berlin - Wannsee.
—— Allen, A.O., and Watson, F.H. (1966). *J. chem. Phys.* **44**, 3431.
Hurst, G.S. and Bortner, T.E. (1959). *Phys. Rev.* **114**, 116.
Hutchinson, G.W. (1948). *Nature, Lond.* **162**, 610.

Jachym, B. (1963). *Acta Phys. polon.* **24**, 243.
Jahnke, J.A., Meyer, L., and Rice, S.A. (1971). *Phys. Rev. A* **3**, 734.
—— Holzwarth, N.A.W., and Rice, S.A. (1972). *Phys. Rev. A* **5**, 463.
Jefferies, M.J. and Mathes, K.N. (1970). *IEEE Trans., Elect. Insul.* (*Inst. elect. electron. Engrs*) **EI-5**, 83.
Jortner, J., Kestner, N.R., Rice, S.A., and Cohen, M.H. (1965). *J. chem. Phys.* **43**, 2614.
Kahan, E. and Morant, M.J. (1963). *Proceedings of the D[illegible]m Conference,* p.
—— —— (1965). *Br. J. appl. Phys.* **16**, 943.
Kao, K.C. (1960). Conference paper 60–84, Winter Meeting, AIEE.
—— (1965). *Nature, Lond.* **208**, 279.
—— Calderwood, J.H. (1965). *Proc. Instn. elect. Engrs.* **112**, 597.
—— Higham, J.B. (1961). *J. electrochem. Soc.* **108**, 522.
—— McMath, J.P.C. (1970). *IEEE Trans., Elect. Insul.* (*Inst. elect. electron. Engrs.*) **EI-5**, 64.
Kawashima, A. (1974). *Cryogenics* **14**, 217.
Keenan, T.F. (1972). *Proceedings of the Dublin Conference*, p. 235.
Keshishev, K.O., Meshov-Deglin, L.P., and Shal'nikov, A.I. (1970). *Sov. Phys. JETP. Lett.* **12**, 160.
Kestner, N.R. and Jortner, J. (1973). *J. chem. Phys.* **59**, 26.
Kilpatrick, M. and Luborski, F.E. (1953). *J. Am. chem. Soc.* **75**, 577.
Kleinheins, G. (1969 *a*). *Phys. Lett. A* **28**, 498.
—— (1969 *b*). *Ber. Bunsen. phys. Chem.* **73**, 1011.
—— (1970). *J. Phys. D, Appl. Phys.* **3**, 75.
Kok, J.A. (1961). *Electrical breakdown of insulating liquids.* Philips Technical Library.

—— Corbey, M.M.G. (1956). *Appl. Sci. Res. Hague B* **6**, 197.
—— —— (1957 *a*). *Appl. Sci. Res. Hague B* **6**, 285.
—— —— (1957 *b*). *Appl. Sci. Res. Hague B* **6**, 449.
—— —— (1958). *Appl. Sci. Res. Hague B* **7**, 257.
Komelkov, V.S. (1962). *Sov. Phys., Tech. Phys.* **6**, 691.
Kopylov, G.N. (1964). *Sov. Phys., Tech. Phys.* **8**, 962.
Krasucki, Z. (1966). *Proc. R. Soc. A* **294**, 393.
—— (1968). *Proceedings of the Grenoble Conference,* p. 311.
—— (1971). *Annual Report of the 1971 Conference on electrical insulation and dielectric phenomena,* p. 96.
—— (1972). *Proceedings of the Dublin Conference,* p. 129.
—— Church, H.F., and Garton, C.G. (1962). *Proceedings of the 19th International Conference on large electric systems,* Vol. 2, paper 138. CIGRE, Paris.
Kraus, C.A. and Fuoss, R.M. (1933). *J. Am. chem. Soc.* **55**, 21.
Kronig, R. and Van de Vooren, A.I. (1942). *Physica* **9**, 139.
Kuper, C.G. (1959). *Phys. Rev.* **122**, 1007.
Lacroix, J.C. and Tobazeon, R. (1972). *Proceedings of the Dublin Conference,* p. 93.
Lebedenko, V.N. and Rodionov, B.U. (1972). *Sov. Phys. JETP Lett.* **16**, 411.
LeBlanc, O.H. (1959). *J. chem. Phys.* **30**, 1443.
Lekner, J. (1967). *Phys. Rev.* **158**, 130.
—— (1968 *a*). *Phys. Lett. A* **27**, 341.
—— (1968 *b*). *Phil. Mag.* **18**, 1281.
Levine, J. and Sanders, T.M. (1962). *Phys. Rev. Lett.* **8**, 159.
—— —— (1967). *Phys. Rev.* **154**, 138.
Lewa, C. (1968). *Acta Phys. polon.* **34**, 165.
Lewis, T.J. (1953). *Proc. Instn. elect. Engrs.* **100**, 2A, 141.
—— (1956). *J. appl. Phys.* **27**, 645.
—— (1957). *Proc. Instn. elect. Engrs.* **104**, B17, 493.
—— (1958). *Br. J. appl. Phys.* **9**, 30.
—— (1959). *Prog. Dielec.* **1**, 97.
—— Ward, B.W. (1962). *Proc. R. Soc. A* **269**, 233.
Liao, T.W. and Anderson, J.G. (1953). *Trans. Am. Inst. elect. Engrs.* **72**, Pt. 1, 641.
Llewellyn Jones, F. (1953). *Rep. Prog. Phys.* **16**, 216.
Low, M.J.D. (1958). *J. electrochem. Soc.* **105**, 103.
Loveland, R.J., Le Comber, P.G., and Spear, W.E. (1972 *a*). *Phys. Rev. B* **6**, 3121.
—— —— —— (1972 *b*). *Phys. Lett. A* **39**, 225.
McClintock, P.V.E. (1969). *Phys. Lett. A* **29**, 1969.
—— (1973). *J. low Temp. Phys.* **11**, 277.
MacFadyen, K.A. (1955). *Br. J. appl. Phys.* **6**, 1.
—— Helliwell, G.C. (1959). *J. electrochem. Soc.* **106**, 1022.
Maksiejewski, J.L. and Tropper, H. (1954). *Proc. Instn. elect. Engrs.* **101**, Pt. II, 183.
Malkin, M.S. and Schultz, H.L. (1951). *Phys. Rev.* **83**, 1051.
Maron, S.H. and Prutton, C.F. (1970). *Principles of physical chemistry*, 4th edn., p. 449. Collier-Macmillan, Int. Edns. London.
Marshall, J.H. (1954). *Rev. scient. Instrum.* **25**, 232.

Massey, H.S.W. and Burhop, E.H.S. (1969). *Electronic and ionic impact phenomena*, 2nd edn., Vol. 1. Clarendon Press, Oxford.

Matuszewski, T., Terlecki, J., and Sulocki, J. (1972). *Proceedings of the Dublin Conference*, p. 189.

Maxwell, J. Clerk. (1892). *A treatise on electricity and magnetism*, Vol. I, 3rd edn., p. 276. Clarendon Press, Oxford.

Meek, J.M. and Craggs, J.D. (1953). *Electrical breakdown of gases*, Clarendon Press, Oxford.

Megahed, I.A.Y. and Tropper, H. (1971). *J. Phys. D, Appl. Phys.* **4**, 446.

Mentalecheta, Y., Delacote, G., and Schott, J. (1966). *C. r. hebd. Séanc. Acad. Sci. Paris* **262**, 892.

Metzmacher, K.D. and Brignell, J.E. (1968 *a*). *J. scient. Instrum.* (*J. Phys. E*) **1**, 1219.

—— —— (1968 *b*). *Proceedings of the Grenoble Conference*, p. 145.

Meyer, L. and Reif, F. (1958). *Phys. Rev.* **110**, 279.

—— —— (1961). *Phys. Rev.* **123**, 727.

—— Davis, H.T., Rice, S.A., and Donnelly, R.J. (1962). *Phys. Rev.* **126**, 1927.

Miller, L.S., Howe, S., and Spear, W.E. (1968). *Phys. Rev.* **166**, 871.

Minday, R.M., Schmidt, L.D., and Davis, H.T. (1969). *J. chem. Phys.* **50**, 1473.

—— —— —— (1971). *J. chem. Phys.* **54**, 3112.

—— —— —— (1972). *J. phys. Chem.* **76**, 442.

Mirza, J.S., Smith, C.W., and Calderwood, J.H. (1970). *J. Phys. D, Appl. Phys.* **3**, 580.

—— —— —— (1972). *Proceedings of the Dublin Conference*, p. 198.

Miyakawa, T. and Dexter, D.L. (1969). *Phys. Rev.* **184**, 166.

Morant, M.J. (1960). *J. electrochem. Soc.* **107**, 671.

—— (1963). *Br. J. appl. Phys.* **14**, 469.

—— (1972). *Proceedings of the Dublin Conference*, p. 109.

Mueller, H. (1941). *J. opt. Soc. Am.* **31**, 286.

Mulliken. R.S. (1950). *J. Am. chem. Soc.* **72**, 600.

Murooka, Y., Nagao, S., and Toriyama, Y. (1961). *Electrotech. J. Jap.* **6**, 56.

—— —— —— (1964). *Br. J. appl. Phys.* **15**, 1585.

Nelson, J.K. and McGrath, P.B. (1972). *J. Phys. D, Appl. Phys.* **5**, 1111.

—— Salvage, B., and Sharpley, W.A. (1971). *Proc. Instn. elect. Engrs.* **118**, 388.

Nikuradse, A. (1934). *Das Flüssige Dielektrikum.* Springer, Berlin.

Northby, J.A. and Sanders, T.M. (1967). *Phys. Rev. Lett.* **18**, 1184.

Nosseir, A. (1973). *IEEE Trans., Elect. Insul.* (*Inst. elect. electron. Engrs.*) **EI-8**, 4.

—— Hawley, R. (1966). *Proc. Instn. elect. Engrs.* **113**, 2, 359.

O'Dwyer, J.J. (1954). *Aust. J. Phys.* **7**, 400.

—— (1973). *The theory of electrical conduction and breakdown in solid dielectrics*, p. 257. Clarendon Press, Oxford.

Onn, D.G., Smetjek, P., and Silver, M. (1974). *J. appl. Phys.* **45**, 119.

Onsager, L.J. (1934). *J. chem. Phys.* **2**, 599.

Ostroumov, G.A. (1954). *Zh. Tekh. Fiz.* **24**, 1915.

—— (1956). *Sov. Phys. JETP.* **3**, 259.

—— (1962). *Sov. Phys. JETP.* **14**, 317.

Palmer, A.W. and House, H. (1972 *a*). *Proceedings of the Dublin Conference*, p. 195.

—— —— (1972 *b*). *J. Phys. D, Appl. Phys.* **5**, 1106.

Pickard, W.F. (1965). *Prog. Dielect.* **6**, 1.

Pitts, E., Terry, C.C., and Willetts, F.W. (1966). *Nature, Lond.* **210**, 295.
Pruett, H.D. and Broida, H.P. (1967). *Phys. Rev.* **164**, 1138.
Rayfield, G.W. and Reif, F. (1964). *Phys. Rev.* **136**, 1194.
Rhodes, G.M. and Brignell, J.E. (1971). *J. Phys. D* **4**, L47.
—— —— (1972). *Proceedings of the Dublin Conference*, p. 116.
Rice, S.A. (1968). *Acc. chem. Res.* **I**, 81.
—— Allnatt, A.R. (1961). *J. chem. Phys.* **34**, 2144.
—— Jortner, J. (1965). *Prog. Dielect.* **6**, 184.
Riegler, A. (1969). *Br. J. appl. Phys.* **2**, 1423.
Robinson, K.A. and Stokes, R.H. (1965). *Electrolyte solutions*, 2nd edn., pp. 129–31. Butterworths, London.
Robinson, M.G. and Freeman, G.R. (1973 *a*). *Can. J. Chem.* **51**, 1010.
—— —— (1973 *b*). *Can. J. Chem.* **51**, 641.
—— —— (1974). *Can. J. Chem.* **52**, 440.
Rose, J. (1967). *Molecular complexes.* Pergamon Press, Oxford.
Rudenko, N.S. and Tsvetkov, V.I. (1966). *Sov. Phys., Tech. Phys.* **10**, 1417.
Sakr, M.M. and Gallagher, T.J. (1964). *Br. J. appl. Phys.* **15**, 647.
Salvage, B. (1951). *Proc. Instn. elect. Engrs.* **98**, 15.
Sano, M. and Akamatsu, H. (1963). *Bull. chem. Soc. Jap.* **36**, 480.
Savoye, E.D. and Anderson, D.E. (1967). *J. appl. Phys.* **38**, 3245.
Saxe, R.F. and Lewis, T.J. (1955). *Br. J. appl. Phys.* **6**, 211.
Scaife, B.K.P. (1973). *Cooperative phenomena*, eds. H.Haken and M.Wagner, pp. 174–82. Springer, Berlin.
Scaife, W.G. (1974). *J. Phys. D, Appl. Phys.* **7**, 647.
Schiller, R. (1972). *J. chem. Phys.* **57**, 2222.
Schmidt, W.F. (1968). *Z. Naturforsch.* **23b**, 126.
—— Allen, A.O. (1968). *J. phys. Chem.* **72**, 3730.
—— —— (1969). *J. chem. Phys.* **50**, 5037.
—— —— (1970 *a*). *J. chem. Phys.* **52**, 4788.
—— —— (1970 *b*). *J. chem. Phys.* **52**, 2345.
—— Schnabel, W. (1971 *a*). *Z. Naturforsch. A* **26**, 169.
—— —— (1971 *b*). *Ber. Bunsen. phys. Chem.* **75**, 654.
Schnabel, W. and Schmidt, W.F. (1973). *J. Polymer Sci. Symp.* No. 42, 273.
Schneider, J.M. and Watson, P.K. (1970). *Phys. Fluids* **13**, 1948.
Schoepe, W. and Rayfield, G.W. (1973). *Phys. Rev.* **7**, A211.
Schottky, W. (1914). *Phys. Z.* **15**, 872.
Schulz, G.J. (1962). *Phys. Rev.* **128**, 178.
Schwarz, K.W. (1970). *Phys. Rev. Lett.* **24**, 648.
—— (1972). *Phys. Rev.* **6**, 837.
—— Stark, R.W. (1969). *Phys. Rev. Lett.* **22**, 1278.
Schynders, H., Meyer, L., and Rice, S.A. (1965). *Phys. Rev. Lett.* **15**, 187.
—— —— —— (1966). *Phys. Rev.* **150**, 127.
Secker, P.E. (1970). *J. Phys. D, Appl. Phys.* **3**, 1073.
—— Hughes, J.F. (1969). *Proc. Instn. elect. Engrs.* **116**, 1785.
—— Lewis, T.J. (1965). *Br. J. appl. Phys.* **16**, 1649.
Sharbaugh, A.H. and Watson, P.K. (1962). *Prog. Dielect.* **4**, 199.
—— —— (1963). Report No. 63-RL-3426C. General Electric Research Laboratories. Schenectady, New York.
—— Cox, E.B., Crowe, R.W., and Auer, P.L. (1955). *Annual Report of the 1955 Conference on electrical insulation and dielectric phenomena,* p. 16.
—— Crowe, R.W., and Cox, E.B. (1956). *J. appl. Phys.* **27**, 806.

Shockley, W. (1951). *Bell System Tech. J.* **30**, 990.
Sibillot, P. and Coelho, R. (1972). *Proceedings of the Dublin Conference*, p. 166.
—— —— (1974). *J. Phys. (Fr.)* **35**, 141.
Silver, M. (1965). *J. chem. Phys.* **42**, 1011.
Simmons, J.G. (1967). *Phys. Rev.* **155**, 657.
Singh, B., Chadband, W.G., Smith, C.W., and Calderwood, J.H. (1972). *J. Phys. D, Appl. Phys.* **5**, 114.
—— Smith, C.W. and Calderwood, J.H. (1972). *Proceedings of the Dublin Conference,* p. 202.
Sletten, A.M. (1959). *Nature, Lond.* **183**, 311.
—— Lewis, T.J. (1963). *Br. J. appl. Phys.* **17**, 883.
Smetjek, P., Silver, M., Dy, K.S., and Onn, D.G. (1973). *J. chem. Phys.* **59**, 1374.
Smith, C.W. and Calderwood, J.H. (1968). *Proceedings of the Grenoble Conference,* p. 161.
Sommer, W.T. (1964). *Phys. Rev. Lett.* **12**, 271.
Spear, W.E. (1969). *J. Non-cryst. Solids* **1**, 197.
Springett, B.E. (1967). *Phys. Rev.* **155**, 139.
Staas, F.A. and Severijns, A.P. (1969). *Cryogenics* **9**, 422.
Steingart, M. and Glaberson, W.I. (1970). *Phys. Rev. A* **2**, 1480.
Stekol'nikov, J.S. and Ushakov, V.Y. (1966). *Sov. Phys., Tech. Phys.* **10**, 1307.
Stokes, C.C. (1845). *Trans. Camb. phil. Soc.* **8**, 287.
Strigel, R. (1934). *Arch. Electrotech.* **28**, 671.
Stuetzer, O.M. (1959). *J. appl. Phys.* **30**, 984.
—— (1960). *J. appl. Phys.* **31**, 136.
—— (1963). *Phys. Fluids.* **6**, 190.
Sugita, K., Sato, T., and Toriyama, Y. (1960). *Br. J. appl. Phys.* **11**, 539.
Swan, D.W. (1961). *Proc. phys. Soc.* **78**, 423.
—— (1962 *a*). *Nature, Lond.* **196**, 977.
—— (1962 *b*). *Br. J. appl. Phys.* **13**, 208.
—— (1963). *Proc. phys. Soc.* **82**, 74.
—— (1964). *Proc. phys. Soc.* **83**, 659.
—— Lewis, T.J. (1960). *J. electrochem. Soc.* **107**, 180.
—— —— (1961). *Proc. phys. Soc.* **78**, 448.
Takeda, S.S., Houser, N.E., and Jarnagin, R.C. (1971). *J. chem. Phys.* **54**, 3195.
Taylor, R.J. and House, H. (1972 *a*). *Proceedings of the Dublin Conference*, p. 1.
—— —— (1972 *b*). *J. Phys. D, Appl. Phys.* **5**, 1465.
Terlecki, J. (1962). *Nature, Lond.* **194**, 172.
Tewari, P.H. and Freeman, G.R. (1968). *J. chem. Phys.* **49**, 4394.
—— —— (1969). *J. chem. Phys.* **51**, 1276.
Thomas, W.R.L. (1972). *Annual Report of the 1972 Conference on electrical insulation and dielectric phenomena*, p. 52.
—— (1973 *a*). Private communication, to be published.
—— (1973 *b*). *Annual Report of the 1973 Conference on electrical insulation and dielectric phenomena,* p. 130.
Thomson, J.J. and Thomson, G.P. (1928). *Conduction of electricity through gases.* 3rd edn., Vol. 1, pp. 193 ff. Cambridge University Press.
Timko, C.A., Penney, G.W., and Osterle, J.F. (1965). *Proc. Instn. elect. Engrs.* **53**, 141.
Toriyama, Y., Sato, T., and Mitsui, H. (1964). *Br. J. appl. Phys.* **15**, 203.
Townsend, J.S. (1910). *The theory of ionization of gases by collision.* Constable, London.

Tropper, H. (1961). *J. electrochem. Soc.* **108**, 144.
Tyndall, A.M. and Powell, C.F. (1930). *Proc. R. Soc. A* **129**, 162.
Vij, J.K. and Scaife, W.G. (1974). Paper presented at 12th Meeting of European High Pressure Group, Marburg. Proceedings not published.
von Hippel, A. (1937). *J. appl. Phys.* **8**, 815.
—— (1946). *Trans. Faraday Soc. A* **42**, 78.
von Laue, M. (1925). *Ann. Phys. Leipzig* **76**, 261.
Walden, P. (1906). *Z. phys. Chem. Leipzig* **55**, 207.
—— Ulich, H. (1923). *Z. phys. Chem. Leipzig* **107**, 219.
Ward, A.L. (1958). *Phys. Rev.* **112**, 1852.
Ward, B.W. (1967). *Seminarium Wyladowania w Dielektrykach Cieklych*, p. 73. Published by Polish Academy of Sciences, Wroclaw 1968. This paper is in English.
—— Lewis, T.J. (1960). *J. electrochem. Soc.* **107**, 191.
—— —— (1963). *Br. J. appl. Phys.* **14**, 368.
Washburn, H.W. (1933). *Phys. Rev.* **4**, 29.
Watson, P.K. (1955). *Annual Report of the 1955 Conference on electrical insulation and dielectric phenomena,* p. 21.
—— Higham, J.B. (1953). *Proc. Instn. elect. Engrs.* **100**, 11A, 163.
—— Sharbaugh, A.H. (1960). *J. electrochem. Soc.* **107**, 516.
—— —— (1963). Report No. 63-RL-3212C. General Electric Research Laboratories, Schenectady, New York.
—— Schneider, J.M., and Till, R.H. (1970). *Phys. Fluids* **13**, 1955.
Weber, K.H. and Endicott, H.S. (1956). *Trans. Am. Inst. elect. Engrs.* **75**, 371.
Weiss, J. (1960). *Nature, Lond.* **186**, 751.
Wilks, J. (1967). *The physical properties of liquid and solid helium.* Clarendon Press, Oxford.
Williams, R.L. (1957). *Can. J. Phys.* **35**, 134.
Willis, W.L. (1966). *Cryogenics*, **6**, 279.
Woolf, M.A. and Rayfield, G.W. (1965). *Phys. Rev. Lett.* **15**, 235.
Zaky, A.A. and Hawley, R. (1973). *Conduction and breakdown in mineral oil.* Peter Peregrinus Ltd., London.
—— Tropper, H., and House, H. (1963). *Br. J. appl. Phys.* **14**, 651.
Zein El-Dine, M.E. and Tropper, H. (1956). *Proc. Instn. elect. Engrs. C* **103**, 35.
—— Zaky, A.A., Hawley, R., and Cullingford, M.C. (1964). *Nature, Lond.* **201**, 1309.
—— —— —— —— (1965). *Proc. Instn. elect. Engrs.* **112**, 580.
Zessoules, N., Brinkerhoff, J., and Thomas, A. (1963). *J. appl. Phys.* **34**, 2010.

# Author Index

# Subject Index